AI for Decision Intelligence in Critical Systems

This book is a multi-disciplinary reference on how domain-aware artificial intelligence (AI) models can outperform generic approaches by addressing sector-specific complexities. It offers comparative frameworks, reproducible case studies, and real-world applications of emerging AI methods.

Collectively, *AI for Decision Intelligence in Critical Systems* emphasizes a unifying theme: the effective deployment of AI to strengthen decision-making, enhance system reliability, and mitigate risks in domains where precision, trust, and efficiency are critical.

This edited volume brings together twenty-one chapters of original research, each exploring how AI, machine learning (ML), and deep learning (DL) are shaping innovation across critical domains. The book highlights the application of advanced architectures—including Convolutional Neural Networks (CNNs), Quaternion Neural Networks (QCNNs), Large Language Models (LLMs), and Gradient-Boosted Decision Trees (GBDTs)—to solve complex, domain-specific challenges.

Concerning computer vision and infrastructure safety, chapters discuss the use of CNNs and QCNNs for automated road crack detection, offering scalable approaches to improving transportation safety while reducing dependence on manual inspections. With regard to software engineering, contributions focus on leveraging ML, DL, and LLMs to enhance software quality assurance, minimize defects, and improve resilience in high-stakes industries. Additional chapters examine ML-driven methods, particularly GBDT, to uncover non-linear drivers of equity valuation across sectors, supporting more accurate forecasts and risk-sensitive decision-making.

Academics and researchers in computer science, AI, and data science, industry professionals in transportation, software engineering, finance, and policymakers seeking to apply AI systems effectively will find this book useful.

Advances in Applied Mathematics

Series Editor: Daniel Zwillinger

An Introduction to Partial Differential Equations with MATLAB®
Matthew P. Coleman and Vladislav Bukshtynov

Handbook of Exact Solutions to Mathematical Equations
Andrei D. Polyanin

Introducing Game Theory and its Applications, Second Edition
Elliott Mendelson and Dan Zwillinger

Modeling Operations Research and Business Analytics
William P. Fox and Robert E. Burks

Decision Analysis through Modeling and Game Theory
William P. Fox

Advances in High-Order Predictive Modelling
Dan Gabriel Cacuci

Introduction to Financial Mathematics, Second Edition
Kevin J. Hastings

General Quantum Variational Calculus
Svetlin G. Georgiev and Khaled Zennir

Advanced Engineering Mathematics with MATLAB®, Sixth Edition
Dean G. Duffy

Exact Methods for Nonlinear PDEs
Andrei D. Polyanin

Markov Processes, Second Edition
James R. Kirkwood

Advances in High-Order Sensitivity Analysis
Dan Gabriel Cacuci

Mining Complex Networks, Second Edition
Bogumił Kamiński, Paweł Prałat, and François Théberge

The Math Thought Mindset
Logic, Reasoning, and Proofs
James L. Burk

Neutrosophic Computing and Vague Data Analysis
Edited by Florentin Smarandache and Zahid Khan

AI for Decision Intelligence in Critical Systems
Edited by Shahab Saquib Sohail, Arpita Soni, Satish Mandavalli, Shantanu Kumar, and Gautam Siddharth Kashyap

https://www.routledge.com/Advances-in-Applied-Mathematics/book-series/CRCADVAPPMTH?pd=published,forthcoming&pg=1&pp=12&so=pub&view=list

AI for Decision Intelligence in Critical Systems

Edited by
Shahab Saquib Sohail, Arpita Soni,
Satish Mandavalli, Shantanu Kumar and
Gautam Siddharth Kashyap

CRC Press
Taylor & Francis Group
Boca Raton London New York

CRC Press is an imprint of the
Taylor & Francis Group, an **informa** business
A CHAPMAN & HALL BOOK

First edition published 2027
by CRC Press
2385 NW Executive Center Drive, Suite 320, Boca Raton FL 33431

and by CRC Press
4 Park Square, Milton Park, Abingdon, Oxon, OX14 4RN

CRC Press is an imprint of Taylor & Francis Group, LLC

ISBN: 978-1-041-30484-5 (hbk)
ISBN: 978-1-041-30483-8 (pbk)
ISBN: 978-1-003-77481-5 (ebk)

DOI: 10.1201/9781003774815

Typeset in CMR10 font
by KnowledgeWorks Global Ltd.

Dedicated to the innovators and guardians of critical systems, whose work with AI helps make our world safer and smarter.

Contents

Editors

Dr. Shahab Saquib Sohail is an assistant professor in the Department of Computer Science and Engineering at Jamia Hamdard, New Delhi. He previously served as a senior assistant professor at VIT Bhopal University. He holds a PhD in computer science from Aligarh Muslim University. He is recognized among the top 5 researchers globally in the Scopus database for work related to ChatGPT and among the top 2% of AI and Computer Vision researchers worldwide (Stanford–Elsevier list). He has authored more than 100 SCI-indexed journal papers, including 70 Q1 and Q2 publications. His research has appeared in high-impact venues such as *Nature Machine Intelligence*, *Information Fusion*, *IEEE Transactions on Big Data*, and *WIREs Data Mining and Knowledge Discovery*, as well as leading conferences including INTERSPEECH, IJCNN, ICASSP, and ICDM workshops. With over 3,000 Google Scholar citations, his research spans computational intelligence, recommender systems, and computational social science. Dr. Sohail is also an active collaborator with international research groups and a dedicated mentor to emerging scholars in AI and machine learning.

Arpita Soni is a senior IT professional with over two decades of experience in software engineering, quality assurance, and program management. She specializes in generative AI, automation, and digital transformation across banking, healthcare, and supply chain sectors. A certified Project Management Professional (PMP) and Certified Scrum Master (CSM), she is also a Senior Member of IEEE and a fellow of the British Computer Society (BCS). Ms. Soni has led large-scale enterprise AI initiatives that enhance operational efficiency, compliance, and reliability. She has authored multiple research papers on AI and machine learning, including work on low-resource chatbots and AI integration into software development lifecycles. She is a frequent keynote speaker, session chair, and reviewer for leading IEEE, Elsevier, IGI Global, and Springer journals and conferences. She is also the author of books such as *AI Sustainability* and *Advanced Statistical Techniques for Data Mining.*

Satish Mandavalli is a software engineer at Microsoft with over twenty years of experience across finance, banking, healthcare, and enterprise IT systems. As a chartered accountant, he uniquely bridges finance and technology to design-intelligent, data-driven solutions. His work focuses on applying AI and machine learning to optimize financial processes, improve risk management, and enable predictive analytics in real-world business environments. Mr. Mandavalli is deeply invested in integrating smart, secure, and scalable AI-driven systems into financial operations. He is a strong advocate for innovation and continues to explore emerging technologies that enhance transparency, efficiency, and decision-making in financial and critical systems.

Shantanu Kumar is a senior software engineer at Amazon, where he has been instrumental in developing and scaling key e-commerce initiatives since 2016. He played a central role in building and expanding Buy with Prime, enabling seamless integrations with Shopify, BigCommerce, Salesforce, and Meta platforms. His expertise spans scalable API design, secure checkout systems, and data-driven orchestration engines which have contributed significantly to Amazon's global commerce ecosystem. He has also worked on machine learning–based recommendation systems for Prime Video and modernized large-scale data ingestion pipelines. Mr. Kumar is a recognized mentor and leader, ranked among the top 1% mentors on the ADP List. He has conducted over 50 professional development sessions worldwide and has guided hundreds of professionals in career growth and technical leadership. A graduate of the National Institute of Technology (NIT) Kurukshetra, Mr. Kumar has received multiple performance awards at Amazon and continues to drive innovation in scalable, intelligent systems.

Gautam Siddharth Kashyap is a PhD researcher at Macquarie University, specializing in the alignment of Large Language Models (LLMs) via the Helpfulness, Harmlessness, and Honesty (HHH) framework. His doctoral research focuses on developing principled methods to align LLMs with human values, leading to publications at EMNLP, EACL, and more. Beyond his doctoral research, Dr. Kashyap also contributes to NLP for social good, focusing on the development of ethical, inclusive, and reliable AI systems.

Contributors

Rafiq Ali
DSEU-Okhla,
New Delhi, India

Amit Anand
National Institute of Technology
Jamshedpur, India

Venu Madhav Aragani
Eliassen Group
Reading, MA USA

Pushkar Arora
DSEU-Okhla
New Delhi, India

Mohammad Anas Azeez
MBUZAI
Abu Dhabi, UAE

Krishna Gandhi
Independent Researcher
Texas, USA

Priyam Ganguly
Widener University
Chester, PA, USA

Ramakrishna Garine
University of North Texas
Denton, TX, USA

Srinivas Raju Gottimukkala
John Deere Financial
Johnston, IA, USA

Vijay Govindarajan
Colorado State University
Fort Collins, CO, USA
Expedia Group Inc.
Seattle, WA, USA

Zohaib Hasan Siddiqui
Jamia Hamdard
New Delhi, India

Niharika Jain
VIPS-GGSIPU
New Delhi, India

Nipun Joshi
Cornell University
Ithaca, NY, USA

John Akkarakaran Jose
Capital One Financial
Chicago, IL, USA

Abdullah Mohammad
DSEU-Okhla
New Delhi, India

Subrat Prasad
Indian Institute of Technology
Roorkee, India

Sushant Kr. Ray
University of Delhi
New Delhi, India

Parth Saxena
Independent Researcher
Houston, TX, USA

Ebad Shabbir
DSEU-Okhla
New Delhi, India

Raj Sonani
Cornell University
Ithaca, NY, USA

Sahil Tripathi
Jamia Hamdard
New Delhi, India

Pankaj Verma
Indian Institute of Management
Bangalore, India

Part I

Introduction

Chapter 1

Introduction: From Generic AI to Domain-Aware Decision Systems

Shahab Saquib Sohail, Arpita Soni, Satish Mandavalli, Shantanu Kumar, and Gautam Siddharth Kashyap

1.1 Introduction

The field of Artificial Intelligence (AI) has been experiencing a swift evolution during the last several decades, changing how data is processed, analyzed, and used to make a decision-making process easier and more efficient [5]. The initial AI systems were mainly developed as a general-purpose tool that could complete general tasks, including classification and prediction and pattern recognition across various fields of use, e.g. [7], prediction [4], and pattern recognition [6]. Although these systems had remarkable computing capabilities, they frequently lacked a sense of contextual background and so could not be well applied in the complex world.

As the availability of domain-specific data grows and the need for reliable automated decision support, both the trend towards creating domain-aware decision systems has strongly changed [11]. Such systems combine expert knowledge with learning algorithms, which allows more accurate, interpretive, and credible results. In this chapter, the author presents the shift in generic AI models towards domain-conscious systems and explains their applicability in intelligent applications today.

1.2 Evolution from Generic AI to Domain-Aware Systems

Conventional AI designs emphasized the creation of generalized architectures which could be used to a broad problem set. Rule-based systems based on

DOI: 10.1201/9781003774815-1

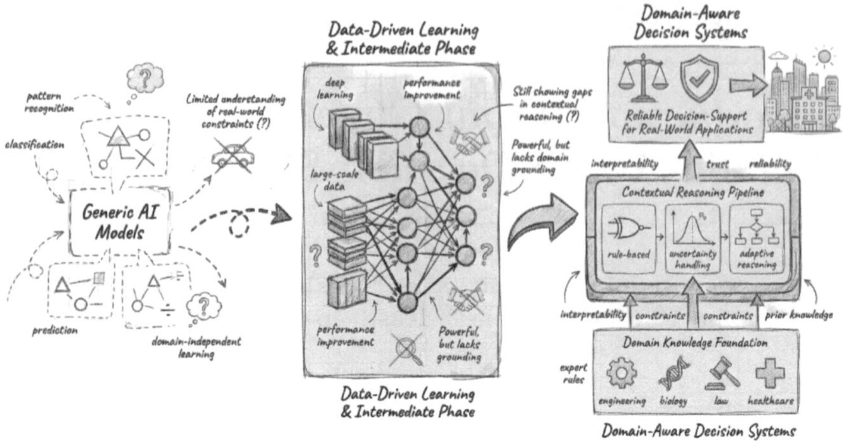

FIGURE 1.1: Evolution from Generic AI Models to Domain-Aware Decision Systems.

the ideas of [8], classical machine learning [2], and early neural networks [10] were created to find patterns in data without paying explicit attention to the characteristics of a domain. Though this was scalable, it frequently did not reflect such subtle contextual relationships of specialized fields.

New AI powers such as the invention of deep learning [3] and massive data processing [14] only enhanced generic AI. Nevertheless, the scholars quickly noted that the boost in performance was not enough in the field of health care, finance, and engineering, which are vital sectors of life that require enhancement through AI-controlled methods and programs [9], finance [13], and engineering [1]. It is against this realization that hybrid strategies of data-driven learning with expert knowledge, ontologies, and domain restrictions were developed. Domain-aware systems are this next developmental move since they instantiate contextual information into model designs and decision chains. These systems modify their reasoning process based on domain specific rules standards and goals which results in more valuable and dependable outputs. Figure 1.1 shows how the general AI models have gradually evolved over time into domain-sensitive decision systems, and the major advancements in technology and methodology.

1.3 Importance of Domain Knowledge in Decision-Making

The introduction of domain knowledge into AI systems is an essential factor in enhancing the quality of the decision and robustness of the system [12].

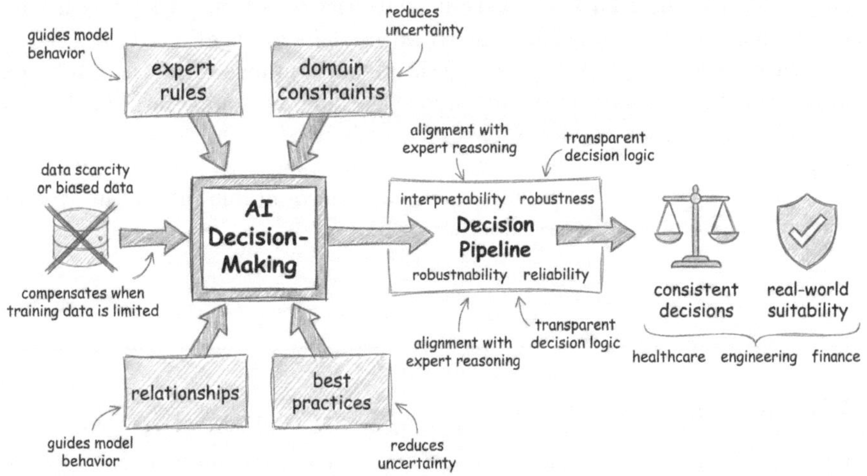

FIGURE 1.2: Role of Domain Knowledge in Enhancing AI Decision-Making.

Domain knowledge contains data on relationships, constraints and best practices in some structured format that is not necessarily visible in raw data. This knowledge when combined with learning algorithms controls model behavior and uncertainty is minimized.

Domain-based decision systems are more interpretative as they match predictions to known expert reasoning. This congruence is especially useful in high stakes environments where there is a need to maintain transparency and accountability. Indicatively, in the medical diagnosis, the use of clinical guidelines assists in ensuring that the recommendations are in accordance with the accepted practise. In addition, the domain knowledge is helpful in the process of generalization in low-data cases. In the case of scarce and/or biased training data, domain-aware models may use previous knowledge to preserve the performance. Consequently, these systems are more robust, flexible and applicable in practical environment. The importance of domain knowledge to build better interpretability, reliability and perform better in decision-making within AI systems is noted in Figure 1.1.

1.4 Conclusion

The shift toward the use of generic AI models by domain-specific decision systems is an indication of the increasing demand of smart solutions that integrate both the efficiency and the understanding of the context. Although generic AI has good basis on learning patterns based on data, it is usually not complex and sensitive enough to be used in applications that need it. Domain-knowledge systems overcome these shortcomings by incorporating expert knowledge, domain constraints and situational reasoning into the decision plan. This integration can be translated to enhanced accuracy, interpretability and reliability. Since AI keeps developing, domain-sensitive solutions are likely to become a key to creating trustworthy and accountable intelligent systems in a wide range of application domains.

Bibliography

[1] Yu Ao, Shaofan Li, and Huiling Duan. Artificial intelligence-aided design (AIAD) for structures and engineering: a state-of-the-art review and future perspectives. *Archives of Computational Methods in Engineering*, pages 1–28, 2025.

[2] Vijay Govindarajan, Pratik Patel, Sahil Tripathi, Md Azizul Hoque, and Gautam Siddharth Kashyap. Magic-enhanced keyword prompting for zero-shot audio captioning with clip models. *arXiv preprint arXiv:2509.12591*, 2025.

[3] Yann LeCun. A path towards autonomous machine intelligence version 0.9. 2, 2022-06-27. *Open Review*, 62(1):1–62, 2022.

[4] Bryan Lim and Stefan Zohren. Time-series forecasting with deep learning: a survey. *Philosophical transactions of the royal society a: mathematical, physical and engineering sciences*, 379(2194), 2021.

[5] Bernard Marr. Artificial intelligence and the future of marketing. *FORBES. Retrieved from https://www. forbes. com/sites/bernardmarr/2022/09/09/artificial-intelligence-and-the-future-of-marketing*, 2022.

[6] Shervin Minaee, Yuri Boykov, Fatih Porikli, Antonio Plaza, Nasser Kehtarnavaz, and Demetri Terzopoulos. Image segmentation using deep learning: A survey. *IEEE transactions on pattern analysis and machine intelligence*, 44(7):3523–3542, 2021.

[7] Shervin Minaee, Nal Kalchbrenner, Erik Cambria, Narjes Nikzad, Meysam Chenaghlu, and Jianfeng Gao. Deep learning–based text classification: a comprehensive review. *ACM computing surveys (CSUR)*, 54(3):1–40, 2021.

[8] Stuart Russell, Peter Norvig, and Artificial Intelligence. A modern approach. *Artificial Intelligence. Prentice-Hall, Egnlewood Cliffs*, 25(27):79–80, 1995.

[9] W Shabeena Shah, Zakaria Elkhwesky, K Mohamed Jasim, Esraa Fayez Youssif Elkhwesky, and Fady Fayez Youssif Elkhwesky. Artificial intelligence in healthcare services: past, present and future research directions. *Review of Managerial Science*, 18(3):941–963, 2024.

[10] Arpita Soni, Sahil Tripathi, Gautam Siddharth Kashyap, Manaswi Kulahara, Mohammad Anas Azeez, Zohaib Hasan Siddiqui, Nipun Joshi, and Jiechao Gao. Can we predict your next move without breaking your privacy? *arXiv preprint arXiv:2507.08843*, 2025.

[11] Roman Lukyanenko Tremblay, Alfred Castillo, and Wolfgang Maass. Domain knowledge in artificial intelligence. *Data & Knowledge Engineering*, page 2, 2025.

[12] Laura Von Rueden, Sebastian Mayer, Katharina Beckh, Bogdan Georgiev, Sven Giesselbach, Raoul Heese, Birgit Kirsch, Julius Pfrommer, Annika Pick, Rajkumar Ramamurthy, et al. Informed machine learning–a taxonomy and survey of integrating prior knowledge into learning systems. *IEEE Transactions on Knowledge and Data Engineering*, 35(1):614–633, 2021.

[13] Yi Yang, Mark Christopher Siy Uy, and Allen Huang. Finbert: A pretrained language model for financial communications. *arXiv preprint arXiv:2006.08097*, 2020.

[14] Daochen Zha, Zaid Pervaiz Bhat, Kwei-Herng Lai, Fan Yang, Zhimeng Jiang, Shaochen Zhong, and Xia Hu. Data-centric artificial intelligence: A survey. *ACM Computing Surveys*, 57(5):1–42, 2025.

Chapter 2

Comparative Analysis of CNN and Quaternion CNN for Image-Based Road Crack Detection in Computer Vision

Parth Saxena

2.1 Introduction

Road maintenance, automated crack detection using Deep Learning (DL) [5, 39]. This is particularly useful because it eliminates the need for manual inspection and increases efficiency. Convolutional Neural Networks have been a dominant method for supervised classification and feature extraction for images [4, 10]. Through the processing of raw image data through multiple layers, they learn complicated patterns as they operate in a hierarchy. One drawback of conventional CNNs is that they are often based on many parameters which may be costly in computation and might lead to overfitting [31, 10, 4, 39, 16]. This problem has led to other architectures such as quaternion-based models. Quaternions are early mathematical structures that extend complex numbers to a four-dimensional space. Quaternions are composed of one real part and three imaginary parts unlike real or complex numbers, and therefore can be used to represent data in multiple dimensions such as color images. This property enables QNNs to generate richer structural information, increasing learning efficiency by allowing for greater information richness [28, 29]. In a QNN, old convolutional and linear layers are replaced by quaternionic ones. These layers work directly in the quaternion domain, and thus do not have redundancy or additional parameters. This reduction in parameters does not affect the model's ability to learn complex features but is a way of avoiding overfitting while maintaining or even improving accuracy. We test QNNs on image classification and segmentation tasks to establish QNN effectiveness. We use three CNN architectures [2]—AlexNet, VGG16, and a custom CNN that are trained on road crack datasets in the classification stage.

DOI: 10.1201/9781003774815-2

Each model is first tested in its conventional form, followed by a modified version where the standard convolutional and linear layers are replaced with quaternionic layers [36, 37, 3, 27, 12, 14, 11, 38, 13, 24, 23]. The performance metrics are recorded and compared, such as accuracy, precision, and parameter efficiency. We segment images using DeepCrack, an DL model for detection of cracks in road surfaces. Just like the classification phase, we alter the core layers to produce a quaternionic version of DeepCrack. This will be used to investigate whether the quaternionic technique improves segmentation accuracy and thereby reduces computational complexity. Thus, this chapter aims to provide a comparative analysis of CNNs and QCNNs for road crack detection.

The chapter is as follows; in the next section, we will look at the theoretical backdrop. Section 2.3 describes the materials and methods. Section 2.4 presents the experimental analysis, and Section 2.5 concludes the chapter with some conclusions and future works.

2.2 Theoretical Background

Neural networks are computational systems that employ the biological neural system. They are composed of many nonlinear computing units called neurons that receive, process and transmit signals in a structured network. The concept of neural networks began with [25] who devised the neuron model with binary thresholds. Yet, [33] created the perceptron, the first form of a neural network capable of learning from linear data in the 1950s. Even with promising potential, the perceptron did not recognize complex patterns, specifically the XOR function, leading to a short-term loss in neural network research until researchers developed Multilayer Perceptrons [9]. MLPs, or feedforward neural networks, introduced multiple layers of perceptron that greatly enhanced their processing abilities. [34] developed backpropagation, a training technique that adjusts network weights to minimize error. This change enabled MLPs to be ubiquitous in solving complex problems. MLPs are also able to approximate nonlinear functions and classify nonlinear data as per single layer perceptrons, thus serving several purposes, including image recognition and classification.

CNNs, particularly useful for visual processing, were also a major development in neural networks. [21] proposed the use of CNNs in 1989 for recognition of zip codes by hand, and it quickly became popular after AlexNet was victorious at the ImageNet Challenge 2012 [20]. CNNs function by convolution, the mathematical algorithm that takes features from images. This reduces the number of learning parameters and makes CNNs computationally efficient. CNNs have multiple layers, including convolutional layers, activation functions, such as ReLU, pooling layers, and fully connected layers. They also use parameter sharing, which enables the same filter to be applied to different parts of the image. The detection of edges and textures allows CNNs to detect these features making them especially useful for image classification [8], object detection [22], and medical image analysis [1]. Based on these architectures, modern DL models can improve in accuracy and efficiency for many applications. CNNs use pooling to provide a summary of localized features in the image. Max pooling, for example, selects the maximum value in a region while other pooling methods average or weight an average by pixels. Pooling is in general advantage because it is in variance translation and therefore it has a way to capture the existence of a feature rather than its exact location. This property is particularly useful when sorting images of different sizes because pooling provides the classification layer with a fixed size input regardless of the size of the input. Dropout is another important technique used by CNNs to avoid overfitting and help generalize models. Overfitting is a process where model learns patterns as well as noise, and thus poor performance on unseen data. Dropout addresses this problem by swiping a small number of neurons to zero during training which changes the network architecture each iteration. This induces robust features for the model that cannot be obtained by relying on specific neurons. Dropout does offer a major advantage in generalization, but with the disadvantage of more training time, which often means two or three times as many iterations for convergence. Dropout rate is the probability of retaining a unit and is generally between 0.5 and 0.8 in hidden layers. A higher dropout rate requires more neurons, while a higher dropout rate may not ensure a sufficient regularization.

With the increasing number of datasets, the performance of DL models becomes more efficient; thus, data augmentation is a commonly practiced method to artificially expand training data. This is particularly useful in the case of classification when the models are to generalize across many different types of input images. Data augmentation is accomplished by application of rotations, random cropping, horizontal and vertical flipping, color jittering, edge enhancement, and PCA adjustments. These changes create more varied training samples, which subsequently enhances the model's ability to recognize features under different conditions.

2.3 Materials and Methods

2.3.1 Quaternion Neural Networks

Quaternions, introduced by [32], are a translation of complex numbers into four dimensions. Each quaternion is composed of a real and three imaginary quaternions that are subject to a certain multiplication. These rules define quaternion algebra, which allows for addition, scalar multiplication, conjugation, and the Hamilton product, a generalization of multiplication for quaternions [17]. Quaternions are widely applied to applications that require three-dimensional transformations, such as computer graphics, physics, and robotics. The use of quaternions has led to an increase in CV and DL in recent years because they can represent and process three-dimensional data better than real-valued methods. Quaternions keep more information, especially in representing colors, which make them useful in image processing. Quaternions are particularly useful in DL for QCNNs. These networks extend traditional CNNs by using quaternion-valued inputs, weights and activation functions. The Hamilton product, in QCNNs, does not use the dot product but can be used in an easier and faster representation of multidimensional data. Plus, the use of weightings between color channels enables greater extraction of features while increasing computational redundancy to achieve better performance for functions such as image classification [18] or recognition of objects [40]. Plus, a quaternion convolution computation computes a quaternion valued filter on an input feature map and produces an output feature map with a structural transformation. In QCNN activation functions may be applied separately to each quaternion part, which makes it possible to combine ordinary activation functions such as ReLU or sigmoid into the quaternion structure. It retains the differences of color channels but it also preserves space and spectral information. This means that quaternion-based models can perform better in imaging restoration [41], classification, and object detection than real-valued CNNs.

2.3.2 Image Segmentation

Image segmentation is a common process used in CV that involves the segmenting of an image into segments or pixels where pixels within a segment have common features such as color, intensity, or texture [15]. Labeling is intended to mark each pixel so that objects or areas of interest in an image are properly separated. Segmentation of images can be used for many purposes, including medical imaging [6], facial recognition [19], and object recognition. Scientists have, over the years, developed a number of different segment algorithms to suit different objectives and needs. Tagging is one of the simplest and most widely used methods of segmenting images, in that you select a threshold value for the intensity of an image. Pixels with intensities below this threshold are put into one category while those with values below this threshold are put into another. This method is especially useful for binary segmentation in which an image is divided into foreground and background [7]. Bilevel thresholding has a single threshold value for two segments, while multi-level thresholding has multiple threshold values for more segments. One of the most well known thresholding techniques is Otsu's [30], which determines the optimal threshold value by minimizing the difference between the two resulting segments. This is accomplished by viewing the image histogram and choosing the threshold that minimizes the inter-segment variance. Unlike global thresholding, adaptive thresholding sets threshold values based on local image properties. This method is especially useful for images with poor lighting. There are several thresholding algorithms available, including

Sauvola's algorithm [26] that adjusts the threshold for each pixel by local mean and standard deviation. Bernsen's method [30], characterized by the threshold being the average of the minimum and maximum intensity values in a neighborhood. Niblack's method [26] takes a threshold with local mean and standard deviation within sliding window to calculate it. Isodata thresholding [26], that iterates the threshold for a given measure of the mean intensity values of pixels above and below a threshold estimate. Tagging methods are computationally efficient but can be problematic with image complexity based on double intensity distributions or noise.

Histogram based segmentation consists of viewing the intensity distribution of images to identify peaks and valleys consistent with objects or regions. It is computationally efficient because the image is processed in a single pass. But one problem is that it can be difficult to distinguish peaks, especially when intensity distributions overlap greatly. Clustering techniques are an effective segmentation approach when dealing with images that have no known objects or regions. Clustering groups pixels that have similar characteristics, usually as a function of color, texture, or intensity. K-Means is a commonly used unsupervised learning algorithm for image segmentation. It splits the pixels into k clusters by decreasing the difference among clusters. [35] introduced FCM clustering, a variant of K-Means. FCM allows pixels to belong to more than one cluster, but at different granularities than K-Means where each pixel is assigned to one cluster. This flexibility makes FCM particularly useful for images with ambiguous boundaries. It is an algorithm that is somewhat like K-Means but includes a fuzzifier parameter which controls the fuzzyness of the clusters. The goal is to minimize an error function by iteratively updating cluster centroids and membership values until convergence. Region-based segmentation, which combines pixels based on spatial continuity and not on pixel types, is another important segmentation technique. The seed points are the first pixels used, chosen according to predefined criteria. It merges the similar pixels from neighbouring pixels into the same segment. Regional growing is a widely used region technique where new pixels are added to a segment if the pixels meet similarity criteria based on intensity, color, or texture. This is an effective method for segmenting the most abstract regions, but it is prone to noise and can be difficult to manipulate widely-simplified images.

2.4 Experimental Analysis

This chapter explores the use of PyTorch, an open-source ML library developed by Facebook AI Research in 2017, for training DL models on crack detection in images. PyTorch provides built-in support for Deep Neural Networks (DNNs) and tensor computations that can leverage GPU acceleration, making it a suitable choice for computationally intensive tasks like image segmentation and classification. This chapter was performed using a 12GB Nvidia Titan Xp GPU, AMD Ryzen 5 1600 Six-Core Processor and 32GB RAM. Training data is comprised of concrete with and without cracks. The data set is split between the two types: positive (images with cracks) and negative (images without cracks). Each category contains 20,000 images, 227 $\times$ 227 $\times$ 3, generated from 458 high resolution images, 4032 $\times$ 3024 pixels. Another secondary dataset of 400 images (448 $\times$ 448$\times$ 3) of asphalt cracks was also used for further testing. A customized CNN was utilized for training, which had four convolutional layers and ReLU activation functions and 2 $\times$ 2 max pooling. Finally, the outputs were passed through a single-layer, a ReLU activation function and a Sigmoid activation function. Training was conducted over 10 epochs, one and ten batches, learning 0.01 per hour. SGD was used as the optimization algorithm and BCELoss was used as the loss function.

2.4.1 Result Analysis

One of the most significant aspects of this chapter was the use of QNNs, which is the representation of the input image as quaternion. Quaternion mathematics reduces complex numbers into space, thereby allowing more compact and efficient representations of data. All layers, which were convolutional and fully connected, were replaced by quaternion versions in the quaternion-based CNN via PyTorch modules including QuaternionLinear and QuaternionConv. Similar changes were applied to AlexNet and VGG16 to investigate the relationship between quaternion models and standard DL architectures. These results show that quaternion-based models required about 75% less parameters than the more traditional models. But this efficiency also meant a greater computational cost, requiring 21% longer training times. Results of performances across multiple datasets indicated that more accurate models were trained with larger datasets.

Tables 2.1 to 2.10 reflect the findings from the analysis using VGG16, AlexNet and the custom CNN for different sizes of data. Results showed that for a dataset with 21,000 or more images, all models performed best in terms of accuracy at 99%. Interestingly, VGG16 performed well on smaller datasets of 7,000 images. Quaternion based models performed better on smaller datasets than the traditional CNNs. They consistently beat their counterparts by over 10% for samples with less than 1,750 images. But, on larger sets of data, such as the Same Parameter CNN, SMCNNs showed better generalization and lower overfitting tendencies. This was further evidenced by testing on the secondary dataset that contains asphalt cracks. Despite this, the normal CNNs generally performed better, especially for datasets of 3,500 to 7,000 images, but quaternion models were overfitting in larger datasets. VGG16 was the most appropriate model for generic crack detection because it proved to be much more accurate even on diverse data.

TABLE 2.1: Average Accuracy (Mean ± Std) of CNN Models Across Different Dataset Sizes Over 10 Epochs

Dataset Size	Accuracy (Mean ± Std)
28K	0.994 ± 0.004
21K	0.990 ± 0.006
14K	0.981 ± 0.025
7K	0.976 ± 0.024
3.5K	0.970 ± 0.024
1.75K	0.917 ± 0.111
0.7K[a]	0.925 ± 0.072
0.35K[b]	0.951 ± 0.011

TABLE 2.2: Average Accuracy (Mean ± Std) of Custom CNN Models Across Different Dataset Sizes Over 10 Epochs

Dataset Size	Accuracy (Mean ± Std)
28K	0.996 ± 0.001
21K	0.992 ± 0.001
14K	0.993 ± 0.002
7K	0.990 ± 0.001
3.5K	0.980 ± 0.004
1.75K	0.972 ± 0.004
0.7K[a]	0.920 ± 0.017
0.35K[b]	0.510 ± 0.000

TABLE 2.3: Average F-Score (Mean ± Std) of CNN Models Across Different Dataset Sizes Over 10 Epochs

Dataset Size	F-score (Mean ± Std)
28K	0.992 ± 0.005
21K	0.988 ± 0.006
14K	0.978 ± 0.026
7K	0.974 ± 0.024
3.5K	0.966 ± 0.025
1.75K	0.887 ± 0.161
0.7K[a]	0.907 ± 0.074
0.35K[b]	0.946 ± 0.014

TABLE 2.4: Average F-score (Mean ± Std) of Custom CNN Models Across Different Dataset Sizes Over 10 Epochs

Dataset Size	F-score (Mean ± Std)
28K	0.994 ± 0.002
21K	0.991 ± 0.001
14K	0.992 ± 0.002
7K	0.988 ± 0.001
3.5K	0.977 ± 0.005
1.75K	0.969 ± 0.007
0.7K[a]	0.906 ± 0.022
0.35K[b]	0.640 ± 0.000

TABLE 2.5: Average Accuracy (Mean ± Std) of CNN Models Across Different Dataset Sizes Over 1 Epoch

Dataset Size	Accuracy (Mean ± Std)
28K	0.904 ± 0.085
21K	0.888 ± 0.087
14K	0.885 ± 0.092
7K	0.866 ± 0.090
3.5K	0.830 ± 0.132
1.75K	0.737 ± 0.156
0.7K[a]	0.818 ± 0.051
0.35K[b]	N/A

TABLE 2.6: Average Accuracy (Mean ± Std) of Custom CNN Models Across Different Dataset Sizes Over 1 Epoch

Dataset Size	Accuracy (Mean ± Std)
28K	0.984 ± 0.003
21K	0.981 ± 0.010
14K	0.966 ± 0.015
7K	0.844 ± 0.155
3.5K	0.795 ± 0.084
1.75K	0.784 ± 0.079
0.7K[a]	0.633 ± 0.171
0.35K[b]	0.500 ± 0.000

TABLE 2.7: Average Accuracy (Mean ± Std) of CNN Models for Asphalt Dataset Over 10 Epochs

Dataset Size	Accuracy (Mean ± Std)
28K	0.724 ± 0.067
21K	0.701 ± 0.082
14K	0.728 ± 0.046
7K	0.776 ± 0.048
3.5K	0.763 ± 0.103
1.75K	0.776 ± 0.067
0.7K[a]	0.763 ± 0.034
0.35K[b]	0.811 ± 0.055

TABLE 2.8: Average Accuracy (Mean ± Std) of Custom CNN Models for Asphalt Dataset Over 10 Epochs

Dataset Size	Accuracy (Mean ± Std)
28K	0.755 ± 0.085
21K	0.771 ± 0.098
14K	0.816 ± 0.065
7K	0.782 ± 0.035
3.5K	0.789 ± 0.073
1.75K	0.753 ± 0.099
0.7K[a]	0.728 ± 0.118
0.35K[b]	0.658 ± 0.159

TABLE 2.9: Average F-Score (Mean ± Std) of CNN Models for Asphalt Dataset Over 10 Ppochs

Dataset Size	F-score (Mean ± Std)
28K	0.579 ± 0.141
21K	0.531 ± 0.239
14K	0.604 ± 0.132
7K	0.661 ± 0.090
3.5K	0.642 ± 0.204
1.75K	0.683 ± 0.142
0.7K[a]	0.717 ± 0.094
0.35K[b]	0.795 ± 0.042

TABLE 2.10: Average F-Score (Mean $\pm$ Std) of Custom CNN Models for Asphalt Dataset Over 10 Epochs

Dataset Size	F-score (Mean $\pm$ Std)
28K	0.634 $\pm$ 0.170
21K	0.656 $\pm$ 0.167
14K	0.731 $\pm$ 0.120
7K	0.696 $\pm$ 0.059
3.5K	0.745 $\pm$ 0.093
1.75K	0.628 $\pm$ 0.216
0.7K[a]	0.572 $\pm$ 0.260
0.35K[b]	0.673 $\pm$ 0.020

TABLE 2.11: Average Performance Metrics (Mean $\pm$ Std) of DeepCrack and Quaternion Models

Metric	Mean $\pm$ Std
bT	0.38 $\pm$ 0.05
G	0.979 $\pm$ 0.009
P	0.767 $\pm$ 0.113
R	0.790 $\pm$ 0.060
F	0.785 $\pm$ 0.086

This chapter also investigated a more advanced architecture, DeepCrack. DeepCrack has 13 layers of convolution, with a batch normalization and a ReLU activation function. The model also contains four max-pooling layers with 2 $\times$ 2 filter size and stride of 2 to downsample feature maps. It also uses side output layers that extract feature maps at different depths of the network, and a final fused layer which adds all these multi-scale features. One of DeepCrack's most interesting features is the use of Guided Filtering, which enables better segmentation accuracy by reducing noise in the output. This strategy improves forecasting by using an earlier side output as a guidance map, and improves the final segmentation results. A quaternion version of DeepCrack was also developed by replacing standard convolutional layers with quaternionic layers. The quaternionic model required significantly less parameters but still maintained competitive performance. But, its general accuracy was slightly below that of the standard DeepCrack model, as shown in Figure 2.1 and Table 2.11.

Quaternion and standard models showed improvements from the lowest to mid-level networks while falling down at higher levels. This is because

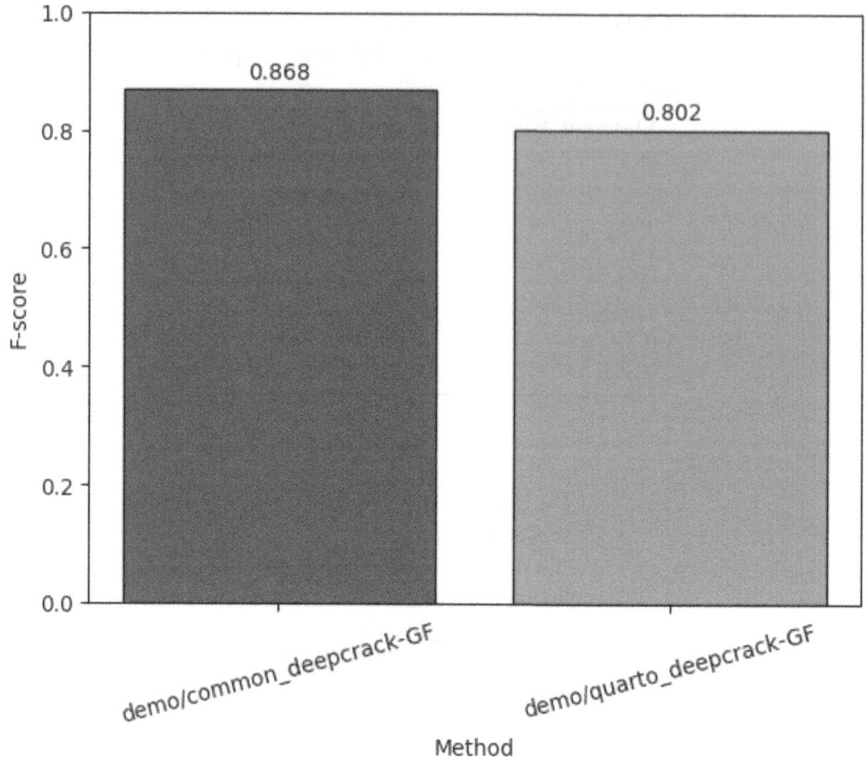

FIGURE 2.1: Precision-Recall Curve of Common-Quaternion DeepCrack.

features at the lower levels were more sensitive to noise than those at the higher layers, capturing more complex features with lower false positives but higher false negatives. The best results were obtained in the fused output layer by combining information from multiple levels.

2.5 Conclusions and Future Works

This chapter analyzes the use of QCNNs and compares them to other models. Quaternions are another promising alternative with comparable or better accuracy and F-scores while significantly reducing the number of parameters. Though the custom CNN used in this chapter was less feature-rich, it performed well. The quaternion-based models had a superior performance to the conventional CNNs in smaller datasets. But, due to their more sophisticated computational process, processing time increased by 21%. The same parameters were used to initialize all models for an even comparison.

Potential improvement include choosing optimal parameters to reduce computational workload and data augmentation to expand the diversity of datasets. For image segmentation tasks such as DeepCrack, quaternion models showed a reduction in parameters but slightly lower performance. Future work could focus on advanced parameter tuning, improved feature merging techniques, and more effective filtering algorithms to refine segmentation accuracy.

Bibliography

[1] DA Agneya, Mogaveera Sahil Shekar, Anagha Bharadwaj, Nandhini Vineeth, and M Lakshmi Neelima. Deep learning in medical image analysis: a survey. In *2024 International Conference on Innovation and Novelty in Engineering and Technology (INNOVA)*, volume 1, pages 1–5. IEEE, 2024.

[2] Laith Alzubaidi, Jinglan Zhang, Amjad J Humaidi, Ayad Al-Dujaili, Ye Duan, Omran Al-Shamma, José Santamaría, Mohammed A Fadhel, Muthana Al-Amidie, and Laith Farhan. Review of deep learning: concepts, cnn architectures, challenges, applications, future directions. *Journal of big Data*, 8(1):53, 2021.

[3] Venu Madhav Aragani. Evaluating reinforcement learning agents for portfolio management. In *2025 Fifth International Conference on Advances in Electrical, Computing, Communication and Sustainable Technologies (ICAECT)*, pages 1–6. IEEE, 2025.

[4] Young-Jin Cha, Wooram Choi, and Oral Büyüköztürk. Deep learning-based crack damage detection using convolutional neural networks. *Computer-Aided Civil and Infrastructure Engineering*, 32(5):361–378, 2017.

[5] Young-Jin Cha, Wooram Choi, Gahyun Suh, Sadegh Mahmoudkhani, and Oral Büyüköztürk. Autonomous structural visual inspection using region-based deep learning for detecting multiple damage types. *Computer-Aided Civil and Infrastructure Engineering*, 33(9):731–747, 2018.

[6] Loan Dao and Ngoc Quoc Ly. A comprehensive study on medical image segmentation using deep neural networks. *arXiv preprint arXiv:2506.04121*, 2025.

[7] Nizam Ud Din, Kamran Javed, Seho Bae, and Juneho Yi. A novel gan-based network for unmasking of masked face. *IEEE Access*, 8:44276–44287, 2020.

[8] Alexey Dosovitskiy. An image is worth 16x16 words: Transformers for image recognition at scale. *arXiv preprint arXiv:2010.11929*, 2020.

[9] Max H Farrell, Tengyuan Liang, and Sanjog Misra. Deep neural networks for estimation and inference. *Econometrica*, 89(1):181–213, 2021.

[10] Yue Fei, Kelvin CP Wang, Allen Zhang, Cheng Chen, Joshua Q Li, Yang Liu, Guangwei Yang, and Baoxian Li. Pixel-level cracking detection on 3d asphalt pavement images through deep-learning-based cracknet-v. *IEEE Transactions on Intelligent Transportation Systems*, 21(1):273–284, 2019.

[11] Srinivas R Gottimukkala. Automating portfolio replication with stock market algorithms. In *2025 8th International Conference on Circuit, Power & Computing Technologies (ICCPCT)*, pages 973–978. IEEE, 2025.

[12] Srinivas R Gottimukkala. A computational approach to replicating correlated portfolios using algorithmic insights from stock market dynamics. In *International Conference on Information and Communication Technology for Intelligent Systems*, pages 265–275. Springer, 2025.

[13] Srinivas R Gottimukkala. Optimizing exotic option pricing: Monte carlo simulation and variance reduction techniques. In *International Conference on Information and Communication Technology for Intelligent Systems*, pages 277–288. Springer, 2025.

[14] Vijay Govindarajan, Pratik Patel, Sahil Tripathi, Md Azizul Hoque, and Gautam Siddharth Kashyap. Magic-enhanced keyword prompting for zero-shot audio captioning with clip models. *arXiv preprint arXiv:2509.12591*, 2025.

[15] Jin Gwan Ha, Hyeonjoon Moon, Jin Tae Kwak, Syed Ibrahim Hassan, Minh Dang, O New Lee, and Han Yong Park. Deep convolutional neural network for classifying fusarium wilt of radish from unmanned aerial vehicles. *Journal of Applied Remote Sensing*, 11(4):042621–042621, 2017.

[16] Hong-wei Huang, Qing-tong Li, and Dong-ming Zhang. Deep learning based image recognition for crack and leakage defects of metro shield tunnel. *Tunnelling and underground space technology*, 77:166–176, 2018.

[17] Teijiro Isokawa, Nobuyuki Matsui, and Haruhiko Nishimura. Quaternionic neural networks: Fundamental properties and applications. In *Complex-valued neural networks: utilizing high-dimensional parameters*, pages 411–439. IGI Global Scientific Publishing, 2009.

[18] Radosław Kycia and Agnieszka Niemczynowicz. 4d hypercomplex-valued neural network in multivariate time series forecasting. *Scientific Reports*, 15(1):23713, 2025.

[19] Lamyanba Laishram, Muhammad Shaheryar, Jong Taek Lee, and Soon Ki Jung. Toward a privacy-preserving face recognition system: A survey of leakages and solutions. *ACM Computing Surveys*, 57(6):1–38, 2025.

[20] Yann LeCun, Yoshua Bengio, and Geoffrey Hinton. Deep learning. *nature*, 521(7553):436–444, 2015.

[21] Grace W Lindsay. Convolutional neural networks as a model of the visual system: Past, present, and future. *Journal of cognitive neuroscience*, 33(10):2017–2031, 2021.

[22] Li Liu, Wanli Ouyang, Xiaogang Wang, Paul Fieguth, Jie Chen, Xinwang Liu, and Matti Pietikäinen. Deep learning for generic object detection: A survey. *International journal of computer vision*, 128(2):261–318, 2020.

[23] Satish Mandavalli. Enhancing dengue outbreak predictions using machine learning: A comparative analysis of models. 2024.

[24] Satish Mandavalli. Enhancing crop image classification: Comparative analysis of augmentation techniques for small datasets. In *International Conference on Information and Communication Technology for Intelligent Systems*, pages 591–604. Springer, 2025.

[25] Warren S McCulloch and Walter Pitts. A logical calculus of the ideas immanent in nervous activity. *The bulletin of mathematical biophysics*, 5(4):115–133, 1943.

[26] Mehrdad Motamed Shariati and Farzane Samiminia. Evaluation of the agreement between the three local thresholding algorithms (Otsu's local thresholding extension, niblack, and sauvola) in choroidal vascularity index measurement using the imagej software. *SN Comprehensive Clinical Medicine*, 7(1):426, 2025.

[27] Lakshmi Narasimha Raju Mudunuri, Praveen Kumar Maroju, and Venu Madhav Aragani. Leveraging NLP-driven sentiment analysis for enhancing decision-making in supply chain management. In *2025 Fifth International Conference on Advances in Electrical, Computing, Communication and Sustainable Technologies (ICAECT)*, pages 1–6. IEEE, 2025.

[28] Titouan Parcollet, Mohamed Morchid, and Georges Linarès. Quaternion convolutional neural networks for heterogeneous image processing. In *ICASSP 2019-2019 IEEE international conference on acoustics, speech and signal processing (ICASSP)*, pages 8514–8518. IEEE, 2019.

[29] Titouan Parcollet, Ying Zhang, Mohamed Morchid, Chiheb Trabelsi, Georges Linarès, Renato De Mori, and Yoshua Bengio. Quaternion convolutional neural networks for end-to-end automatic speech recognition. *arXiv preprint arXiv:1806.07789*, 2018.

[30] Tapaswini Pattnaik and Priyadarshi Kanungo. Non-uniform illumination image thresholding based on k-means clustering and mlr. In *2025 International Conference on Cognitive, Green and Ubiquitous Computing (IC-CGU)*, pages 1–5. IEEE, 2025.

[31] Yupeng Ren, Jisheng Huang, Zhiyou Hong, Wei Lu, Jun Yin, Lejun Zou, and Xiaohua Shen. Image-based concrete crack detection in tunnels using deep fully convolutional networks. *Construction and Building Materials*, 234:117367, 2020.

[32] Wolf-Dieter Richter. Quaternions without imaginarities or the vector representation of quaternions. *Preprints. org, doi*, 10, 2025.

[33] Frank Rosenblatt. The perceptron: a probabilistic model for information storage and organization in the brain. *Psychological review*, 65(6):386, 1958.

[34] David E Rumelhart, Geoffrey E Hinton, and Ronald J Williams. Learning representations by back-propagating errors. *nature*, 323(6088):533–536, 1986.

[35] Manas Singh Sakrwar, AS Ranadive, and Dragan Pamucar. A novel gustafson–kessel based clustering algorithm using n-pythagorean fuzzy sets. *Systems and Soft Computing*, page 200345, 2025.

[36] Parth Saxena. Evaluating machine learning algorithms for effective credit card fraud detection. In *2025 10th International Conference on Signal Processing and Communication (ICSC)*, pages 375–380. IEEE, 2025.

[37] Parth Saxena and Janak Devashish. Sentiment analysis and topic modeling of patient reviews on psychiatric medications using cloud-based architecture. In *2025 International Conference on Computational, Communication and Information Technology (ICCCIT)*, pages 892–897. IEEE, 2025.

[38] Arpita Soni, Sahil Tripathi, Gautam Siddharth Kashyap, Manaswi Kulahara, Mohammad Anas Azeez, Zohaib Hasan Siddiqui, Nipun Joshi, and Jiechao Gao. Can we predict your next move without breaking your privacy? *arXiv preprint arXiv:2507.08843*, 2025.

[39] Allen Zhang, Kelvin CP Wang, Baoxian Li, Enhui Yang, Xianxing Dai, Yi Peng, Yue Fei, Yang Liu, Joshua Q Li, and Cheng Chen. Automated pixel-level pavement crack detection on 3d asphalt surfaces using a deep-learning network. *Computer-Aided Civil and Infrastructure Engineering*, 32(10):805–819, 2017.

[40] Dandan Zhang, Chuan Lin, and Yongcai Pan. A lightweight small object detection network inspired by the visual area v2. *Computers and Electrical Engineering*, 126:110471, 2025.

[41] Jinping Zou, Huan Ren, and Xiang Wang. Quaternion matrix function representation for color image recovery. *Numerical Algorithms*, pages 1–29, 2025.

Chapter 3

Self-Structuring Neural Networks for Optimized Data Analytics Using EMANN-Like Algorithms

Priyam Ganguly and Ramakrishna Garine

3.1 Introduction

The term Deep Learning also refers to (DL) [15], ML is a broad area in a specialized sub-area of Machine Learning (ML) [1], And it is another branch of AI, called Artificial Intelligence [20, 6, 14, 31]. They are often viewed as the cutting-edge in AI. DL is a specific set of ML methods that can be applied to many real-world problems. ML in general is the development of algorithms to learn and improve tasks from an external source without programming. These algorithms find patterns in data and improve decision-making based on past performance. Ultimately, ML is an attempt to create self-improved systems with the minimal human intervention. DL is designed around NNs, computer models of the human brain structure and function. These networks are composed of thousands and thousands of neurons that communicate across neurons and process information in a hierarchy.

A key challenge of DL is understanding the optimal topology of a neural network–how many layers and how many neurons should be placed in each layer. In most modern applications, software engineers manually design network topology using trial and error [21]. This is often a complex experiment with multiple configurations tested to find the best solution for a given problem. But automating DL architecture design would be of great benefit to both speed and adaptability, since humans would no longer spend time or effort on fine-tuning networks. While DL has gained some traction over recent years as a result of growth of companies like Google, Netflix and Amazon, DL has been the foundation of work in the field since the 1990s [19]. Several new ML algorithms with neural networks were developed at this stage, many of which could

DOI: 10.1201/9781003774815-3

already be considered early forms of DL methods. One of these algorithms is Evolving Modular Architecture for Neural Networks (EMANN), developed at the IRIDIA laboratories at the Université Libre de Bruxelles [29]. EMANN was the first incremental method of structuring neural networks, which aimed to automate it. Rather than specifying an architecture, EMANN shifted the topology of the network in the context of training by adding or removing layers and neurons per selected criteria. This flexibility allowed the algorithm to create the optimal network structure for a task.

Therefore, the present analysis focuses on the EMANN algorithm under a microscope of emerging DL trends and technologies. The chapter uses the open source TensorFlow library from Google to implement modernized EMANN algorithms, also known as EMANN-like algorithms. The new algorithms are tested against a benchmark image classification task using the CIFAR-10 [13]. Since this dataset contains 60,000 images representing ten classes, CIFAR-10 makes an ideal benchmark for evaluation of DL models. This chapter aims to compare different EMANN-like algorithms for classification accuracy and to understand how various parameter values affect network performance. The results of the synthesis of these neural network topologies provide an opportunity to determine the effect that differences in number of layers and neurons have on the accuracy of fully connected neural networks when applied to CIFAR-10.

The chapter is as follows; in the next section, we will look at the theoretical backdrop. Section 3.3 describes the materials and methods. Section 3.4 describes the experimental analysis, and Section 3.5 concludes the chapter with some conclusions and future work.

3.2 Theoretical Background

AI is broadly defined as any computer program, algorithm or robot that can be programmed to be intelligent in a given environment. Scholars over the years have come up with differing conceptions of AI's purpose, some aiming for intelligent computing and some building agents that interact with the surroundings to gain maximum benefit from the process. But three main reasons for AI's rapid growth in recent years have been increased computational power (especially on GPUs), DL algorithms increasing, and massive quantities of data. It is projected that AI will be a major contributor to the global economy, with forecasts calling for €12.8 trillion of growth by 2030. The increasing use of AI tools to compete is making the governments in nations around the world, especially Asia, North America, and some European countries look to make money from AI. AI technologies have already transformed multiple industries such as healthcare, finance, transport, and manufacturing. In spite of AI's growing importance, few people have learned the basic concepts. Many people use AI, ML and DL interchangeably, but both have different meanings. ML is a subset of AI that allows systems to learn patterns from data without explicit programming, and DL is a specialized ML oriented on ANN. ML methods can be divided into supervised learning (reading algorithms from labels) or unsupervised learning (learning patterns without labels). Other techniques, such as semi-supervised and reinforcement learning, provide a little more flexibility in the AI model training [22, 23, 2, 18, 10, 8, 9, 25, 11, 17, 16].

DL has gained particular interest because it has been used to solve complex problems like image recognition and natural language processing [4, 5, 28, 27, 12, 26]. ANN are the structural and functional components of DL that mimic the human brain. Neural networks are made up of interconnected artificial neurons grouped in layers to help the neural networks process information more efficiently. As computational power has increased, DL models have become particularly effective in identifying highly complex patterns. One of the key facets of DL is the selection of an activation function which is responsible for how neurons process inputs. Linear activation functions, such as identity function, are rarely used in modern applications. Rather, non-linear activation functions such as the sigmoid function and Rectified Line Unit (ReLU) are commonly used. ReLU specifically has revolutionised DL, helping to train DNNs with ease [30, 7]. ReLU is still the most commonly used activation function of neural networks for simplicity and effectiveness. But, it has "the dying ReLU problem" where negative inputs mean zero output and hinder learning in some cases. The solution to this is Leaky ReLU that allows small nonzero gradients for negative inputs [24]. In addition, ReLU is not restricted to output in probability calculations and requires alternatives, such as the softmax function that binds the output value to one in the classification analysis.

The Multi-Layer Perceptron (MLP) is one of the most basic models of neural networks with all neurons being tied to each other in a single feed forward layer [3]. MLPs use the backpropagation algorithm, which tries to minimize error by deriving a gradient. This means that the error is computed at the output layer and returned to the network to adjust weights and biases so that the network can learn from data effectively. The most challenging task of designing a neural network is to determine the best configuration, such as the number of hidden layers, the number of neurons per layer and learning parameters. Too small networks may not adequately capture complexities in the data, or are too large networks may not apply generalization correctly in the training data, overfitting. These are traditionally trials and error decisions, but automated processes have begun to gain momentum to simplify this process.

3.2.1 EMANN Algorithm

The EMANN algorithm in 1994, a self-structuring neural network classifier that incrementally builds and trains neural networks. Unlike traditional neural networks, which are based on fixed topologies, EMANN changes structure constantly by adding and pruning neurons and modules during training. It enables the network to improve its architecture and learn from data, making it particularly useful for more complex classification tasks. EMANN adopts a positive perspective, which means it doesn't define the topology of its networks first but builds it up gradually. This method ensures that the structure of the network adapts to the problem at hand, increasing efficiency in learning and reducing unnecessary complexity. The most striking feature of EMANN is that its structure is modular; each module is composed of an entirely connected MLP with one hidden layer. The sigmoid function is part of the activation function used throughout the network and is a central parameter in determining the Connection Strength (CS) of neurons–an important factor for network evolution.

EMANN network is a sequential network of modules that are added continuously. Each module complements the others to ensure the robustness of the network learning process. Each module maintains its fixed output layer and the input layer varies from the output of earlier modules, i.e., global network output. This prevents the spread of errors in one module to the other, and thus allows the network to be more resilient. Within each module, neurons are added dynamically during training, following a well-defined process that includes three primary stages:

1. **Ascension Stage** – New neurons are gradually introduced to the hidden layer.
2. **Improvement Stage** – The optimal number of neurons is determined.
3. **Pruning and Recovery** – Unnecessary neurons are removed to maintain efficiency.

This stepwise construction is necessary to refine the network structure with high accuracy. But, one of the most interesting features of EMANN is its CS measurement of the importance of individual neurons in the hidden layer. The CS is the relative sum of the absolute values of its weights linked in a neuron that represents the overall influence of a neuron in the network. This CS value is important because it is related to the steepness of the sigmoid activation function. A neuron with higher CS values is more likely to be a part of the learning process of the network. EMANN classifies neurons on the basis of their CS values. Ascension is the beginning of a new module's training process. The hidden layer of the module consists first of one neuron, and then of an additional set of neurons. Neuron addition has a fixed AS, meaning that at every AS training period, an additional neuron is added. This stage goes on until at least one neuron in the module passes the "settled" state, that is, its CS exceeds the Settling Threshold (ST). Initially, there is an alternative AS value assigned to the first module to avoid the "copy problem", whereby a new module simply copy the outputs of the previous module rather than engaging individually in the learning process. Once at least one neuron settles, the module moves into the improvement stage. This is the phase during which the network waits for a set number of training periods referred to as the Patience Parameter (PP). If another neuron settles during PP epochs, it is added and the waiting time reset. If no more neurons settle, the module moves to the next stage. This stage aims to better code the number of neurons in the sub-layer so that the module can better perform classification without becoming too complex. The module then goes through the Improvement phase where the network determines the usefulness of its neurons in pruning. If the network's accuracy declines after pruning, there is a recovery phase where the module is then trained again to restore the pre-pruning accuracy. Otherwise the module is retained in its pruned form and the algorithm proceeds to construct the next module. The EMANN algorithm adds modules incrementally until the network is at an optimal level. This design ensures that each new module complements previous ones rather than simply refines outputs. The process continues until the MSE improvement between multiple modules reaches a threshold of An Improvement Threshold (IT). At this point, the network has been declared fully constructed and optimized for classification.

3.3 Materials and Methods

EMANN was implemented for this chapter using Python and TensorFlow. TensorFlow, an open-source library developed by Google Brain team, was selected because of its ease of use with calculations in numerical units and compatibility with both GPU and CPU resources. The low level architecture of TensorFlow, while useful, creates highly personalized neural networks, imposes complexity in even the most basic tasks. But, its flexibility made it a useful candidate for the development of self-structuring neural networks that are a crucial part of EMANN. Another important part of this chapter was to identify a benchmark task that best describes EMANN's application in TensorFlow. We chose the CIFAR-10 dataset after careful consideration. It consists of 60,000 color images that are each (32×32) pixels and divided into 10 mutually exclusive categories: airplane, automobile, bird, cat, deer, dog, frog, horse, ship, and truck. The total dataset is split into 50,000 training images divided into five training batches and 10,000 test images, each class represented equally in the test set.

The EMANN algorithm was implemented using three distinct Python files. The first, `run_EMANN_algo.py`, is the main executable file containing the EMANN algorithm, including its key stages: ascension, improvement, pruning, and recovery. The second, `EMANN_original.py`, defines the module class, including methods for adding and pruning neurons, computing

CS, executing backpropagation, and measuring model accuracy. The third, `data_helpers.py` is responsible for preparing and managing CIFAR-10 data batches. One of the main challenges during implementation was enabling automatic modification of the MLP's topology—allowing neurons to be dynamically added or pruned while preserving their weights. Early on, it was decided to represent MLP operations as matrix computations while removing bias terms to simplify calculations. In TensorFlow, data structures are referred to as tensors, and computations are executed as a flow of operations (hence the library's name). The key to implementing EMANN was leveraging matrix operations. For a given hidden layer λ with m neurons, where the preceding layer $\lambda - 1$ had n neurons, the weighted sum z_λ was computed using matrix multiplication. These z_λ values were then passed through the layer's activation function to compute the final outputs. Weight matrices were stored as lists of 1D tensors instead of 2D tensors to help dynamic adaption of the network structure. This allowed for quick addition and removal of neurons by modifying Python lists, which is not supported explicitly in TensorFlow. Such tensor lists were efficiently managed through forward propagation and backpropagation by means of stacking and unstacking.

There have been several modifications to the original EMANN algorithm described in that are now suitable for modern DL requirements. In deeper layers, the CS thresholds for settled" or useless" neurons were different than in the first layer, which led to different threshold values for the first module relative to the subsequent modules. The EMANN model, in contrast to the old model, which had checks for neuron status every decade, added a check threshold parameter as a way to reduce computational overhead. Parameter for an increase in units was implemented to add several neurons at a fixed frequency during the ascension process. The modules were also stacked differently; rather than counting all the outputs of previous modules into new modules, the new arrangement used only the hidden layer of the previous module, akin to a traditional MLP structure. In addition, the number of modules to construct was no longer tied to MSE reduction and hence model complexity could be controlled more finely. An ascension limit was introduced to control for excessive neuron addition prior to progress to the improvement stage. A major goal of this chapter was to develop new incremental learning algorithms using EMANN including modern DL techniques as illustrated in Figure 3.1. To facilitate this, the module class was designed as a separate Python file, allowing different variations to be tested without modifying the main algorithm. Three EMANN-like models were developed: `EMANN_original.py`, which implemented sigmoid activation for all layers and closely resembled the original EMANN; `EMANN_softmax.py`, which utilized softmax activation for all layers; and `EMANN_ReLU.py`, which applied Leaky ReLU ($\alpha = 0.2$) in hidden layers and softmax in the output layer. These variations allowed for experimentation with different activation functions to improve performance on CIFAR-10.

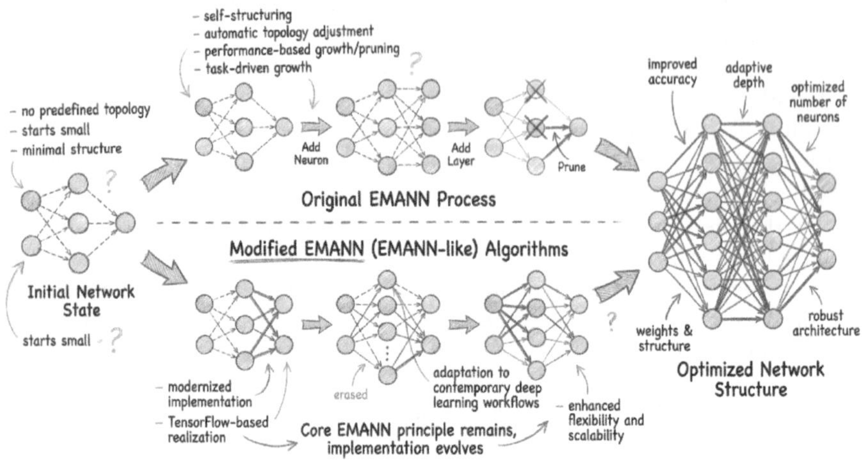

FIGURE 3.1: Progressive Addition of Modules in the Original and Modified EMANN Algorithms.

3.4 Experimental Analysis

The EMANN took place on a Linux Ubuntu 16.04 Bash console running Windows 10 Home 64-bit. The purpose of these tests was to assess the performance of different EMANN-like implementations and examine the influence of several key parameters. To evaluate the influence of specific parameters on model performance, six tests were conducted. These tests focused primarily on the accuracy, network structure, and training behaviour for the three implemented EMANN versions: EMANN_original (sigmoid activation), EMANN_softmax (softmax activation), and EMANN_ReLU (Leaky ReLU activation). Similar analyses were conducted with respect to accuracy under a range of conditions, neuron counts, and behavior trends in CS. To ensure that the results are reliable, statistical tests were performed, such as ANOVA and one-way tests.

The first test examined the influence of ST and AT on accuracy for all three EMANN variants. Both the last training batch and the test set were measured in accuracy at two key points: (1) after the Ascension step and (2) after a full module construction cycle. This test provided insight into the impact these factors had on EMANN's learning experience and final classification performance.

The second test examined the role of different levels of ST and AT on the total number of neurons in the hidden layer. Previous data indicated that neuron count declined marginally after the Ascension stage, and measurements were only made at the end of the complete EMANN cycle. This test was necessary to understand the structure of the network and how efficient EMANN's self-organizing mechanism.

The third test revealed an interesting feature: neurons in both sigmoid and softmax implementations always had the maximum CS and never exceeded the ST threshold, but the Leaky ReLU implementation struggled to do so. Instead, in the ReLU model, the CS maximum had three phases: (1) an increase, (2) a random value of stagnation, and (3) a gradual decline. To further investigate this, the ReLU model was tested 10 times with both ST and AT values to find the number of cases where neurons failed to settle. This information also helped explain this behavior and how changes in ST and AT affected the modeling's ability to adapt well. The fourth test examined the effects of patience parameter on accuracy for the three implementations. As this parameter did not get relevant until after ascension, accuracy measurements were done at the end of the complete EMANN cycle. This test was vital to determining whether adjustments in patience values could improve model stability and learning efficiency in later phases of network construction. The fifth test examined the effects of decreasing the initial number of neurons and increasing the number of neuron during the 'Ascension' stage on accuracy. In that EMANN automatically adds and prunes neurons, a better understanding of the impact of these starting values on learning outcomes was necessary to improve model performance. Results from these tests provided useful information about how to improve network initialization strategies for better accuracy and stability. The sixth test evaluated the performance of the second module whose inputs were from the top-secret layer of the first module. This test was based on the best-performing module configurations from previous tests. The goal was to determine whether stacking multiple EMANN modules could improve the accuracy and learning rate as well.

For the first four tests, ANOVA was conducted one-way or two-way depending on the test. Each combination of parameters was tested three times, and a 5% (0.05) probability confidence level was used for statistical significance. For the last two tests, iterative testing was more straightforward, having multiple iterations of the test. To contextualize EMANN's performance, all accuracy discussions were placed in a 'reference test", a simple MLP model with a single hidden layer and a fixed number of neurons. These tests provided some interesting information about EMANN's behavior and performance. First, threshold sensitivity was significant, since ST and AT had important effects on accuracy and network structure. The sigmoid and softmax implementations were well adapted, but the ReLU implementation was inconsistent with growth in CS. Second, the neuron growth and stability analysis indicated that the number of neurons survived most of the ascent stage with most structural changes occurring at early training stages. Third, there was an unusual confidence score problem for the ReLU implementation, since its neurons did not stabilise and some of these consequences negatively impacted accuracy. Fourth, the patience parameter was related to training accuracy in

TABLE 3.1: Overall Average Accuracy (Mean ± Std) Across Different Sample Sizes for Ascension and Full EMANN Cycle

Samples	Overall Mean ± Std
200	0.035 ± 0.197
500	0.023 ± 0.153
800	0.024 ± 0.156

later stages indicating that a tuned value of patience might improve network refinement. Fifth, network initialization strategies have been found to have a major impact on performance, in particular the decision to select the initial unit count and incremental neuron increases. Finally, the success of the second module evaluation suggested that multi-module EMANN architectures might offer an interesting frontier of future research.

3.4.1 Result Analysis

The reference test, a prebuilt single-layer MLP, was trained using a learning rate of 10^{-3}. However, for EMANN implementations, this setting proved too coarse, so the learning rate was adjusted to 10^{-4} to facilitate convergence. The reference test used an MLP model with 120 hidden neurons and 2000 training epochs. This model achieved an average training accuracy of approximately 60% on the CIFAR-10 dataset, while its testing accuracy was around 46%. These results served as a baseline to compare EMANN's performance. Tables 3.1–3.10 display the test results. The first test investigated how different ST and AT values influenced accuracy. A single EMANN module was used, with a Uselessness Threshold (UT) of 0.013 and a PP of 4000 epochs. The ST values tested were 0.014, 0.016, and 0.019, while the AT values tested were 200, 500, and 800 epochs. Statistical analysis using ANOVA showed that variations in both ST and AT were significant, as well as their interaction. This test showed the average of lower accuracy levels than the reference test, with the highest accuracy (36.2%) being at ST = 0.019 and AT = 200. Importantly, no evidence of overfitting was observed because training and testing accuracy were parallel. The second test addressed the effect of ST and AT on neuron count. The same configurations as the first test were used, and statistical analysis showed that only slight differences were found between sigmoid and softmax implementations. Higher neuron counts were found to be associated with higher accuracy, especially at ST = 0.019 and AT = 200. This indicates that larger-scale networks can capture more detailed patterns in the data and thus improve the classification performance. The third test examined a problem in the ReLU-based EMANN implementation where neurons do not settle well during ascension.

TABLE 3.2: Overall Average Probability Values (Mean ± Std) for Different Thresholds and Interactions

Probability Type	Overall Mean ± Std
Ascension Threshold	0.0039 ± 0.0063
Settling Threshold	0.0033 ± 0.0051
Interaction Probability	0.0047 ± 0.0057

TABLE 3.3: Overall Average Accuracy (Mean ± Std) for Different Thresholds in the Ascension Stage Only and Full EMANN Cycle via Softmax

Setting	Overall Mean ± Std
200–0.014	0.225 ± 0.059
200–0.019	0.299 ± 0.007
500–0.014	0.169 ± 0.031
500–0.019	0.256 ± 0.012
800–0.014	0.172 ± 0.039
800–0.019	0.245 ± 0.022

TABLE 3.4: Overall Average Probability Values (Mean ± Std) for Different Threshold and Interaction Types via Softmax

Probability Type	Overall Mean ± Std
Ascension Threshold	0.147 ± 0.045
Settling Threshold	0.066 ± 0.101
Interaction	0.809 ± 0.093

With the same settings as the previous tests, this chapter examined how different ST and AT values impacted this failure rate. ANOVA found that the settling threshold had a borderline statistical significance (6.62%), and the ascension threshold was much more strongly associated (95.60%). CS quickly increased, stagnated, and then slowened down for ReLU modules. The failure to settle appeared to be linked to higher ST values, as neurons struggled to reach a stable state before CS stagnation. The fourth test examined how the PP affected accuracy. Given previous findings, the AT was set to 200, and the ST was set to 0.019 for sigmoid and softmax modules, and 0.014 for ReLU modules. The tested PP values were 2000, 4000, and 8000. Statistical analysis confirmed that the sigmoid and softmax implementations achieved significantly higher accuracies than the ReLU implementation. The best results were observed in the sigmoid-based module, which sometimes reached

TABLE 3.5: Overall Average Accuracy (Mean ± Std) for Different Thresholds via ReLU

Setting	Overall Mean ± Std
200–0.014	0.147 ± 0.013
200–0.016	0.162 ± 0.021
200–0.019	0.100 ± 0.022
500–0.014	0.102 ± 0.035
500–0.016	0.114 ± 0.012
500–0.019	0.133 ± 0.014
800–0.014	0.124 ± 0.031
800–0.016	0.104 ± 0.040
800–0.019	0.124 ± 0.033

TABLE 3.6: Overall Average Probability Accuracy (Mean ± Std) via ReLU

Probability Type	Overall Mean ± Std
Ascension Threshold	0.251 ± 0.126
Settling Threshold	0.737 ± 0.246
Interaction	0.151 ± 0.191

TABLE 3.7: Overall Performance Comparison of Activation Functions (Mean ± Std)

Activation Function	Overall Mean ± Std
ReLU	3.96 ± 1.15
Sigmoid	3.67 ± 1.96
Softmax	10.74 ± 7.29

TABLE 3.8: Overall Average Probability Values (Mean ± Std) Across Activation Functions

Probability Type	Overall Mean ± Std
Ascension Threshold	0.327 ± 0.462
Settling Threshold	0.147 ± 0.207
Interaction	0.283 ± 0.285

39% accuracy on training batches and 35% on the test set. A slight increase in accuracy was observed with higher patience values, though ANOVA suggested that this correlation was not statistically significant. The underlying explanation could be that longer patience allows more neurons to settle, indirectly leading to a larger network and improved accuracy. The fifth test

TABLE 3.9: Overall Performance of Activation Functions (Mean ± Std)

Activation Function	Overall Mean ± Std
ReLU	0.147 ± 0.010
Sigmoid	0.360 ± 0.017
Softmax	0.269 ± 0.018

TABLE 3.10: Overall Probability Comparison of Different Parameters (Mean ± Std)

Parameter	Overall Mean ± Std
Patience	0.783 ± 0.006
Module Implementation	$1.54 \times 10^{-6} \pm 1.44 \times 10^{-6}$
Interaction	0.871 ± 0.053

assessed the effect of increasing the initial neuron count and unit addition during the Ascension stage. Tests were performed on the sigmoid and ReLU implementations, using ST = 0.019 for sigmoid and both ST = 0.014 and ST = 0.019 for ReLU. Results indicated that increasing the number of initial neurons and the rate of neuron additions led to rapid accuracy improvements. In the sigmoid implementation, when the neuron counts quickly reached 100 (or started near that value), training accuracy sometimes surpassed 50%, with testing accuracy reaching 40%. But, only 27% of the training and 21% of testing improvement in accuracy were achieved with neurons still struggling to settle in the ReLU implementation. But, one major negative was that bigger numbers of neurones increased the computational time of the algorithm, likely due to computational inefficiency with TensorFlow tensors in Python lists. The sixth test was a on how topology effects a second EMANN module. In general, the second module was also accurately estimated to be around 10% accurate equivalent to random guessing in a 10-class classification task. But, when the first module was counted well and was very accurate, the second module improved, achieving up to 20% accuracy in some cases. These results suggest that EMANN's modular design can potentially be DL, but further refinements need to be made so that the second module can achieve accuracy comparable or better than the first module.

3.5 Conclusions and Future Work

The results of the experiments on EMANN show that although it is important to refine the algorithm for complex classification tasks such as CIFAR-10, some important insights have emerged. For the reason that EMANN relies more on stronger connections than does ReLU, it is better to activate softmax and sigmoid-like activation functions than ReLU. Plus, a high settling threshold and low ascension threshold resulted in more neurons in a module; the more neurons, the better is the accuracy due to better pattern memorization. On top of that, measures of increase in neuron count, such as the newly proposed "initial unit number" and "unit increase" were also positive.

Future studies should be directed toward examining the effects of these parameters on deeper modules, other activation functions such as tanh, and self-structuring CNNs, which are based on EMANN. Tests with data such as MNIST could also help to make this a more transparent conclusion. These refinements could also improve EMANN's adaptability and performance in real-world applications.

Bibliography

[1] Pulkit Agrawal, Rajat Arya, Aanchal Bindal, Sandeep Bhatia, Anupriya Gagneja, Joseph Godlewski, Yucheng Low, Timothy Muss, Mudit Manu Paliwal, Sethu Raman, et al. Data platform for machine learning. In *Proceedings of the 2019 international conference on management of data*, pages 1803–1816, 2019.

[2] Venu Madhav Aragani. Evaluating reinforcement learning agents for portfolio management. In *2025 Fifth International Conference on Advances in Electrical, Computing, Communication and Sustainable Technologies (ICAECT)*, pages 1–6. IEEE, 2025.

[3] Fatiha Bouaziz, Daoud Boutana, and Patrick Siarry. Efficient hybrid system based on multi-layer feed-forward neural network with particle swarm optimization algorithm for cardiac arrhythmias classification. *Circuits, Systems, and Signal Processing*, pages 1–25, 2025.

[4] Peng Cui, Zheyan Shen, Sheng Li, Liuyi Yao, Yaliang Li, Zhixuan Chu, and Jing Gao. Causal inference meets machine learning. In *Proceedings of the 26th ACM SIGKDD international conference on knowledge discovery & data mining*, pages 3527–3528, 2020.

[5] Xin Luna Dong, Seungwhan Moon, Yifan Ethan Xu, Kshitiz Malik, and Zhou Yu. Towards next-generation intelligent assistants leveraging llm techniques. In *Proceedings of the 29th ACM SIGKDD Conference on Knowledge Discovery and Data Mining*, pages 5792–5793, 2023.

[6] Mohamed I El Desouki and Wael H Gomaa. Exploring the recent trends of paraphrase detection. *International Journal of Computer Applications*, 975(8887), 2019.

[7] Fabio Galbusera, Gloria Casaroli, and Tito Bassani. Artificial intelligence and machine learning in spine research. *JOR spine*, 2(1):e1044, 2019.

[8] Srinivas R Gottimukkala. Applying the multifractal model of asset returns (MMAR) to financial markets: Insights and limitations. In *2025 10th International Conference on Signal Processing and Communication (ICSC)*, pages 277–282. IEEE, 2025.

[9] Srinivas R Gottimukkala. Automating portfolio replication with stock market algorithms. In *2025 8th International Conference on Circuit, Power & Computing Technologies (ICCPCT)*, pages 973–978. IEEE, 2025.

[10] Srinivas R Gottimukkala. A computational approach to replicating correlated portfolios using algorithmic insights from stock market dynamics. In *International Conference on Information and Communication Technology for Intelligent Systems*, pages 265–275. Springer, 2025.

[11] Vijay Govindarajan, Pratik Patel, Sahil Tripathi, Md Azizul Hoque, and Gautam Siddharth Kashyap. Magic-enhanced keyword prompting for zero-shot audio captioning with clip models. *arXiv preprint arXiv:2509.12591*, 2025.

[12] Nipuni Hewage and Dulani Meedeniya. Machine learning operations: A survey on mlops tool support. *arXiv preprint arXiv:2202.10169*, 2022.

[13] Vikas Kumar and Ayan Rajput. Cifar-10 object detection using. In *Proceedings of International Conference on Recent Trends in Computing: ICRTC 2024, Volume 2*, volume 885, page 313. Springer Nature, 2025.

[14] Soonil Kwon et al. Att-net: Enhanced emotion recognition system using lightweight self-attention module. *Applied Soft Computing*, 102:107101, 2021.

[15] Yann LeCun, Yoshua Bengio, and Geoffrey Hinton. Deep learning. *nature*, 521(7553):436–444, 2015.

[16] Satish Mandavalli. Enhancing dengue outbreak predictions using machine learning: A comparative analysis of models. 2024.

[17] Satish Mandavalli. Enhancing crop image classification: Comparative analysis of augmentation techniques for small datasets. In *International Conference on Information and Communication Technology for Intelligent Systems*, pages 591–604. Springer, 2025.

[18] Lakshmi Narasimha Raju Mudunuri, Praveen Kumar Maroju, and Venu Madhav Aragani. Leveraging nlp-driven sentiment analysis for enhancing decision-making in supply chain management. In *2025 Fifth International Conference on Advances in Electrical, Computing, Communication and Sustainable Technologies (ICAECT)*, pages 1–6. IEEE, 2025.

[19] Omid Panahi. Deep learning in diagnostics. *Journal of Medical Discoveries*, 2(1):1–6, 2025.

[20] Biswal Rashi, YS Kumar Biswal, N Rao, and D Ramchandra. An ai-based customer relationship management framework for business applications. *International Journal of Intelligent Systems and Applications in Engineering*, 12(12s):686–695, 2024.

[21] Sasan Salmani Pour Avval, Nathan D Eskue, Roger M Groves, and Vahid Yaghoubi. Systematic review on neural architecture search. *Artificial Intelligence Review*, 58(3):73, 2025.

[22] Parth Saxena. Evaluating machine learning algorithms for effective credit card fraud detection. In *2025 10th International Conference on Signal Processing and Communication (ICSC)*, pages 375–380. IEEE, 2025.

[23] Parth Saxena and Janak Devashish. Sentiment analysis and topic modeling of patient reviews on psychiatric medications using cloud-based architecture. In *2025 International Conference on Computational, Communication and Information Technology (ICCCIT)*, pages 892–897. IEEE, 2025.

[24] Ashish Kumar Singh and Anoj Kumar. Multi-objective: hybrid particle swarm optimization with firefly algorithm for feature selection with leaky relu. *Discover Artificial Intelligence*, 5(1):192, 2025.

[25] Arpita Soni, Sahil Tripathi, Gautam Siddharth Kashyap, Manaswi Kulahara, Mohammad Anas Azeez, Zohaib Hasan Siddiqui, Nipun Joshi, and Jiechao Gao. Can we predict your next move without breaking your privacy? *arXiv preprint arXiv:2507.08843*, 2025.

[26] Nikko Ström. Scalable distributed dnn training using commodity gpu cloud computing. 2015.

[27] Ermal Toto, ML Tlachac, and Elke A Rundensteiner. Audibert: A deep transfer learning multimodal classification framework for depression screening. In *Proceedings of the 30th ACM international conference on information & knowledge management*, pages 4145–4154, 2021.

[28] Tomihisa Welsh, Michael Ashikhmin, and Klaus Mueller. Transferring color to greyscale images. In *Proceedings of the 29th annual conference on Computer graphics and interactive techniques*, pages 277–280, 2002.

[29] Xingfu Wu, John R Tramm, Jeffrey Larson, John-Luke Navarro, Prasanna Balaprakash, Brice Videau, Michael Kruse, Paul Hovland, Valerie Taylor, and Mary Hall. Integrating ytopt and libensemble to autotune openmc. *The International Journal of High Performance Computing Applications*, 39(1):79–103, 2025.

[30] Pingyue Zhang, Mengyue Wu, Heinrich Dinkel, and Kai Yu. Depa: Self-supervised audio embedding for depression detection. In *Proceedings of the 29th ACM international conference on multimedia*, pages 135–143, 2021.

[31] Nanyang Zhu, Xu Liu, Ziqian Liu, Kai Hu, Yingkuan Wang, Jinglu Tan, Min Huang, Qibing Zhu, Xunsheng Ji, Yongnian Jiang, et al. Deep learning for smart agriculture: Concepts, tools, applications, and opportunities. *International Journal of Agricultural and Biological Engineering*, 11(4):32–44, 2018.

Chapter 4

Scalable Data Science Workflow in the Cloud

Amit Anand and Subrat Prasad

4.1 Introduction

In an increasingly data-driven world, many organizations rely on data science for valuable insight and informed decisions. As data grows, scaling these efforts becomes a concern [22, 23, 16]. Smaller systems are more likely to require more storage space, more computational power, and higher maintenance than their on-premise counterparts, which may not be suitable for large-scale data processing. Cloud-based information enables you to store and edit many thousands of data, keep track of vital details, and write new ideas without spending too much money. This guide explains in a brief but readable manner how the cloud can assist with large-scale data tasks. It discusses what to avoid, why it is useful and helpful tips. Topics include cloud services, design decisions, and data management tips and how to get better information. Many industries use examples to illustrate the benefits of cloud-based solutions in terms of adaptability and productivity. These findings provide a roadmap to the use of the cloud at each stage of a data driven project—from plan to delivery. The goal is therefore to provide clear directions on how to use the power of the cloud to expand the data science activities from the ground up.

The chapter is as follows; in the next section, we will look at the theoretical backdrop. Section 4.3 describes the materials and methods. Section 4.4 describes the challenges, and Section 4.5 concludes the chapter with some conclusions.

DOI: 10.1201/9781003774815-4

4.2 Theoretical Background

Cloud-based data science has brought a fresh approach to how organizations gather, store, and learn from information. With the help of cloud services, even the most extensive datasets can be handled smoothly, Machine Learning (ML) models can be up and running at scale, and teams across the globe can work together without missing a beat [6, 20]. Instead of pouring money into on-site hardware, companies can channel their energy into uncovering insights from their data, making financially responsible decisions that lead to cost-effectiveness. Therefore, this section looks at the key ideas behind cloud-based data science i.e. the different ways to use the cloud, such as data storage, data processing, and ML model deployment. We also explore the core components of cloud-based data science, including different cloud computing paradigms, key benefits, and common challenges in adoption.

4.2.1 Cloud Computing Paradigms

Cloud computing offers a variety of service types to cater to different data science workflow needs. The models are designed to be controlled, user-friendly and scaleable. These models consist of Infrastructure as a Service (IaaS), Platform as a Service (PaaS), and Software as a Service (SaaS), which have very high level of control, scalability and ease of use [4, 5, 3].

4.2.1.1 Infrastructure as a Service

IaaS provides internet-based computing, which enables data science teams to create custom configurations without using physically situated servers. IaaS enables storage, network and computing power, as well as the control of the operating systems and software. Examples of IaaS are Amazon EC2, Google Compute Engine, and Microsoft Azure Virtual Machines. IaaS provides data science services with powerful GPU and TPU instances for demanding ML tasks. This allows for easier configuration of individual machine systems for specialized data processing and model training. Integrates with distributed computing tools such as Apache Spark, Ray and Dask for large-scale data projects. Full control over infrastructure making it the ideal choice for custom, high-performance data science workloads. Organizations can scale resources on demand without incurring the capital costs of in-house hardware Figure 4.1 illustrates the architecture of IaaS-based data science systems.

4.2.1.2 Platform as a Service

PaaS provides a setting that takes care of intruder operations, allowing data scientists to spend less time building models and working with data. These

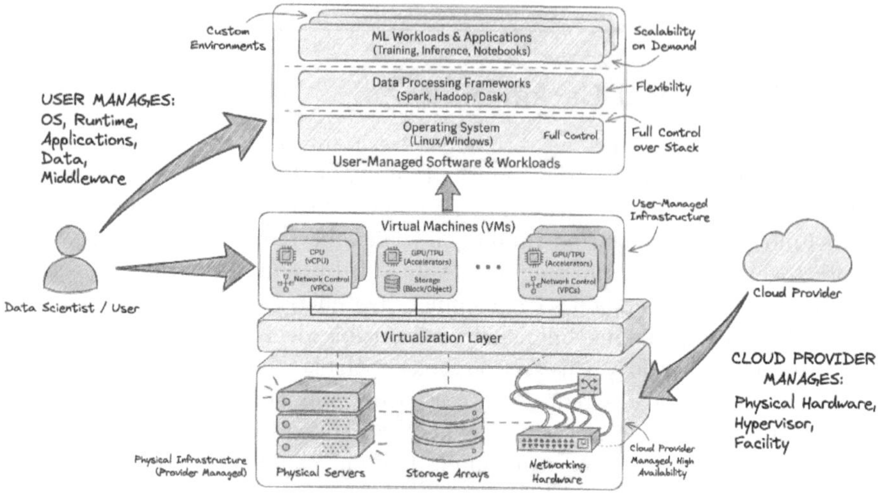

FIGURE 4.1: Architecture of IaaS for Data Science Applications.

platforms often have embedded ML and big data tools that can be used immediately to get started. PaaS services include Google AI Platform, Amazon SageMaker, and Azure Machine Learning. PaaS supports data science via built-in tools for data preprocessing, model training, deployment, and monitoring. Includes Automated Machine Learning (AutoML) capabilities for non-expert users. Provides seamless integration with AI frameworks like TensorFlow, PyTorch, and Scikit-learn. Reduces the burden of managing infrastructure, accelerates AI adoption, and improves collaboration among teams [17, 18, 1, 15, 8, 7, 10, 21, 9, 14, 13]. The architectural representation of PaaS-based data science platforms is shown in Figure 4.2.

4.2.1.3 Software as a Service

SaaS is a cloud-based approach that provides ready-to-use applications for performing tasks. Because these tools run online, there is no need to worry about maintaining infrastructure-related work. The service provider manages all the technical details behind the scenes, leaving the focus on extracting insights. Generally, these applications can be accessed through a web browser or APIs, making it easy for teams to collaborate and fit the tools into existing workflows. SaaS services include Google Colab, IBM Watson Studio, DataRobot, H2O.ai. SaaS supports data science via pre-configured AI/ML environments with minimal setup. Offers drag-and-drop interfaces for model building, making AI accessible to non-technical users. Enables real-time collaboration and sharing of notebooks and models. Simplifies AI adoption for

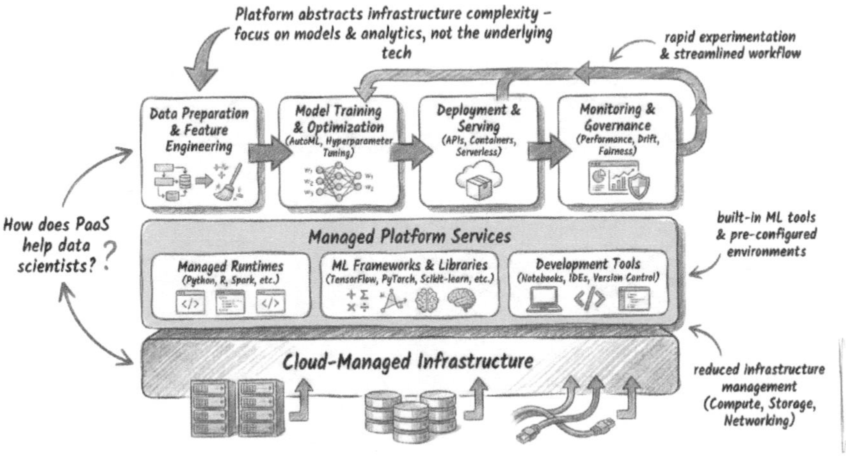

FIGURE 4.2: Architecture of PaaS for Data Science and Machine Learning.

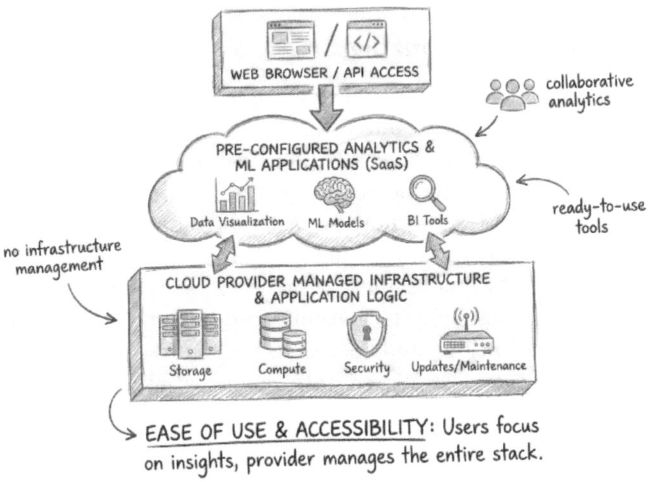

FIGURE 4.3: Architecture of SaaS for Cloud-Based Data Analytics.

organizations by eliminating the need for infrastructure management. The system architecture of SaaS-based data science solutions is presented in Figure 4.3.

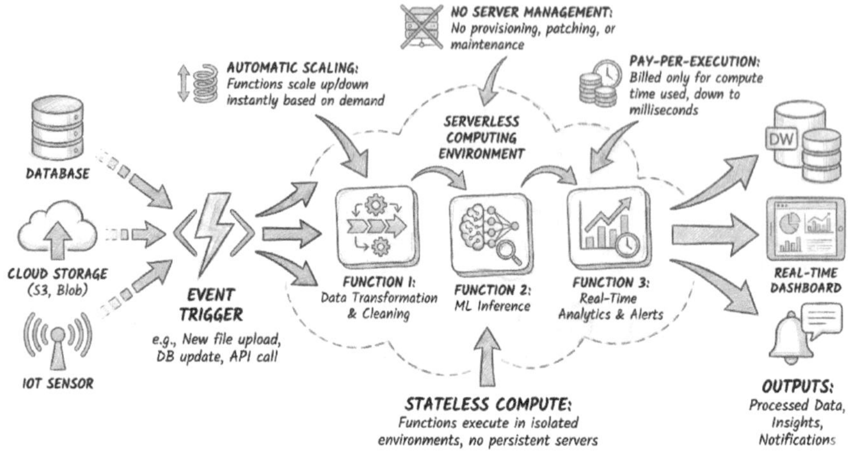

FIGURE 4.4: Architecture of Serverless Computing for Event-Driven Data Science Workflows.

4.2.2 Serverless Computing

Serverless computing allows data science applications to run without managing the underlying servers. Resources are automatically allocated as needed, optimizing cost and performance [2]. For instance, AWS Lambda, Google Cloud Functions, Azure Functions. Serverless computing supports data science via executing data transformation tasks, ML inference, and real-time analytics without provisioning infrastructure. Supports event-driven workflows, such as triggering an AI model inference when new data arrives. Reduces idle resource costs by charging only for actual compute time. Ideal for lightweight, event-driven AI applications and real-time processing workloads. The architectural workflow of serverless-based data science applications is illustrated in Figure 4.4.

4.3 Materials and Methods

One of the biggest challenges when working with data is handling growth in terms of data volume and complexity, processing demands, and evolving business needs. A good data science workflow should be able to handle increasing workloads, multiple data sources and support the continuous development and deployment of models without creating bottlenecks. Imagine we run a busy restaurant. On a typical day, the kitchen staff takes up frequent requests; all dishes are served on time and efficient. But what happens on a night where twice as many customers walk in? It is a well-organized kitchen—chefs are better at making sure their food is ready and extra hands are available when needed. No chaos, just smooth and efficient service. The same concept can be applied to scalable data science workflows. When processing millions of customer transactions, data from the sensor, or a huge amount of image processing, then the system should be able to expand and contract any time, and perform at the same level of need. Just as a restaurant adapts to the rushing crowds of customers, a well-defined data workflow can survive new challenges without ebbing itself of joy. Scalability is not simply a switch to more computing power; it's a means of organizing the workflow in ways that make sense. This includes choosing appropriate methods for data storage, enabling data processing, improving the training of models, and efficiently deploying models in order to make predictions without wasted resource.

4.3.1 End-to-End Workflow Components

Generally, a scalable data science workflow has five elements, each of which is essential to the smooth flow of data processing, model development, and deployment. Raw data is initially collected across sources like relational databases, APIs, IoT sensors and streaming platforms [24, 11]. Storage solution should be responsive to changing data demands, security and availability. Dedicated relational databases are available in the form of Google BigQuery, Amazon RDS, and Azure SQL Database. Large repository platforms like Amazon S3, Google Cloud Storage, and Azure Blob Storage are good places to store text, images, videos, and logs. Apache Kafka, AWS Kinesis, and Google Pub/Sub all support high-throughput ingestion for dynamic analysis and AI. Raw data needs to be cleaned, transformed and arranged prior to analysis so that we can gain valuable insights and have a good training of models. Cloud-based solutions provide managed services to automate preprocessing and scale up data transformation. Cloud-based services such as Apache Spark on Databricks, AWS EMR, and Google Cloud Dataproc enable parallelized transformations of large scale. Batch and streaming data processing is done through services like AWS Glue, Google Dataflow, and Azure Data Factory. A feature store, such as Amazon SageMaker Feature Store, Google Vertex AI Feature Store, and Azure ML Feature Store are useful stores of features.

Scalability in model training is essential for large data sets, deep learning workloads, and computationally intensive experiments. Cloud providers offer specialized AI training environments that support distributed computing and automated optimization. Cloud-based training providers, such as AWS SageMaker, Google Vertex AI and Azure Machine Learning, provide fully managed training streams. Deep learning workloads can be accelerated by using high-performance AWS EC2 P3, Google Cloud TPUs, and Azure ND-series VMs.

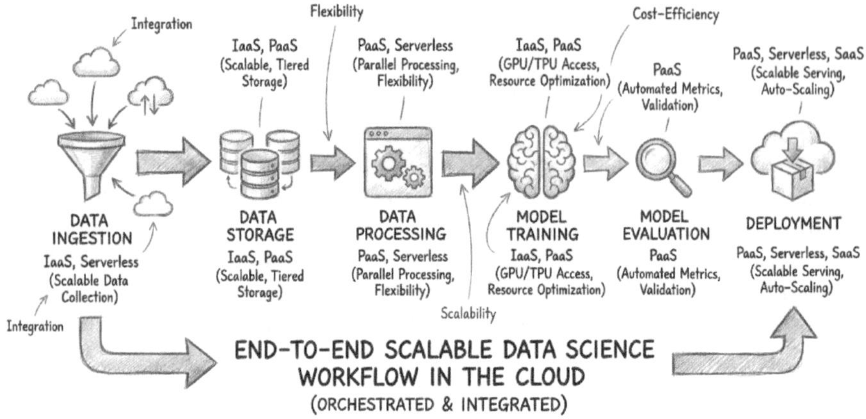

FIGURE 4.5: End-to-End Architecture of a Scalable Data Science Workflow.

Frameworks like Horovod, TensorFlow MirroredStrategy, and Ray allow for parallelized training across multiple nodes. AWS SageMaker Autopilot, Google AI Platform Hyperparameter Tuning, and Azure AutoML improve model performance with very little human intervention. After training, the model is to be easily deployed for real-time or batch inference. Cloud services also support serverless, containerized, and API-based deployments as an option to dynamically scale inference workloads. AWS Lambda, Google Cloud Functions, and Azure Functions provide small-scale, on-demand execution. Kubernetes-based solutions such as AWS EKS, Google GKE and Azure AKS allow for more flexible and scalable AI model hosting. Apache Spark MLlib, AWS Batch, and Google Dataflow are well suited for large batch predictions. Model deployment is simplified through auto-scaling with AWS SageMaker Endpoints, Google AI Predictions, and Azure ML Managed Online Endpoints. Figure 4.5 plots the end-to-end structure of the scalable data science workflow from data ingestion, preprocessing, modeling training, deployment, and inference.

4.3.2 Cloud-Native Technologies

Cloud-Native AI platforms are designed to take advantage of the scalability and flexibility of cloud computing, allowing data scientists to build, deploy, and manage AI models more efficiently. These platforms provide a suite of tools and services that enable data scientists to focus on building and training AI models, rather than worrying about the underlying infrastructure. With cloud-native AI platforms, data scientists can quickly spin up and down resources as needed, reducing costs and increasing productivity. For instance, Azure's

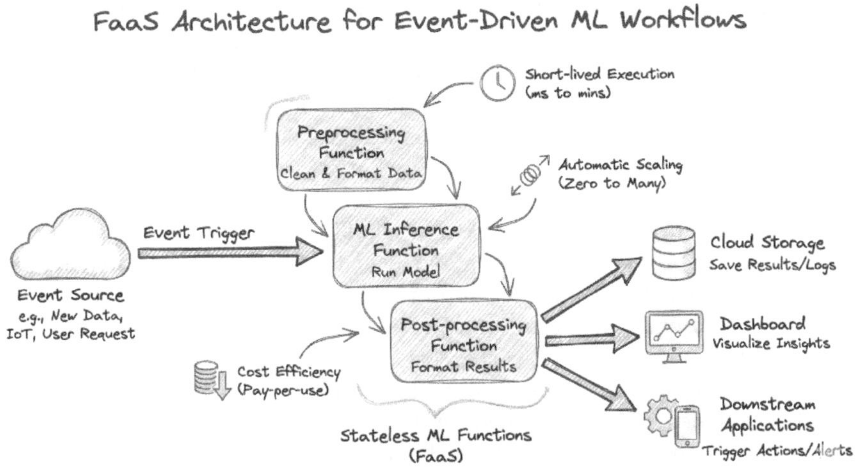

FIGURE 4.6: Architecture of FaaS for Event-Driven Machine Learning Workflows.

cloud-based platform provides a scalable and secure environment for data scientists to build and deploy AI models.

4.3.2.1 Functions-as-a-Service

Functions-as-a-Service (FaaS) is a serverless execution model where ML workloads can be run as small, event-driven functions. These functions can be called upon the presence of a specific event, such as a new data upload, model request, or batch job. For example, AWS Lambda performs ML inference for low latency real-time applications. Google Cloud Functions automates feature engineering or triggers retraining of models as new data arrives, and Azure Functions handles actual time processing of data and flexible AI workflows. Figure 4.6 shows the architecture of FaaS for event-driven machine learning workflows.

4.4 Challenges

Cloud computing has redefined data science by providing scale, flexibility and cost-efficiency allowing firms to process large amounts of data, train ML models and run AI driven applications without a physical infrastructure. But, living in a cloud environment poses its own set of risks that can have an impact on performance, costs, reproducibility and security. Scalability, cost-effectiveness, reproducibility of models and security are all critical components for organizations that want to achieve optimal cloud based data science workflows that are simple and smooth. In order to overcome such challenges, plan ahead, improve resources, and create best practices that meet the specific needs of cloud-based AI and big data workloads. In data science, large datasets, complicated computations, and real-time analytics are often the norm. Without this optimization, performance can slow down and lead to increased costs, longer processing time and inefficiency. Cloud computing gives businesses access to compute resources on-demand, which is very flexible to scale up or down as needed. But, improper use of resources, deficient provisioning, and overused instances can lead to unexpected cost, resulting in massive budgets. To effectively manage cloud costs, cost reduction needs to be managed with resource planning, pricing decision-making, and proactive monitoring. Reproducibility is also a critical element of data science because experiments, data, and models can be replicated for validation, debugging, and continuous improvement. But this also creates a problem in cloud environments because it is hard to track model versions and dependencies. The storage and processing of sensitive data in the cloud poses security risks, including data breaches, unauthorized access, and regulatory compliance issues. International standards are essential.

4.4.1 Case Studies

Fraud is a serious problem for banks because of the millions of transactions they process every day. Traditional rules do not change with a changing fraud pattern that generates false positives and missed fraud activities. AWS pipelines enable financial services to deploy real-time, scalable and intelligent fraud detection through ML and event driven architectures. AWS provides a "servicioless, cost-efficient, and fast" ecosystem that can analyze transaction data, anomalies, and immediate action. JPMorgan Chase faced challenges when it came to fraud detection, because of high false positives, limited scalability, and poor adaptation to new fraud patterns, as well as compliance requirements [12]. They implemented a hybrid cloud approach, using private cloud for sensitive data and AWS for scaleable computing, and using multiple regions and containerized microservices to modernize. Their streaming-first data pipeline utilized Kafka for ingestion in real-time, Spark Streaming for feature computation, and Redis for low-latency caching. Google has won web search from using scalable data science algorithms. Google ranks web pages and groups them by scale in order to index the ever growing internet. The firm can also process large datasets simultaneously using parallel processing algorithms such as MapReduce [19]. This parallelization not only accelerates the indexing process, but also ensures that the search engine remains responsive to the evolving nature of the web. Thus, the worldwide search engine still provides users with quick, relevant and comprehensive results. Figure 4.7

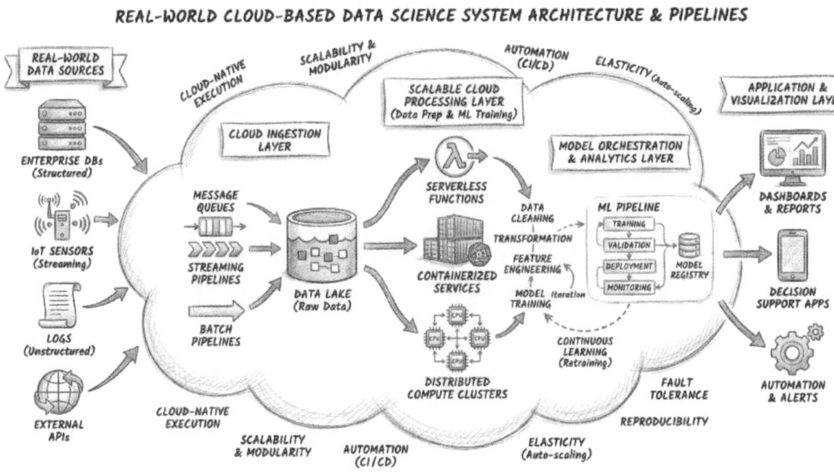

FIGURE 4.7: System Architecture and Data Processing Pipelines for Real-World Case Studies.

shows the types of system architectures and pipelines that are used in the presented real world case studies.

4.5 Conclusions

Cloud-based scaled data science workflows are a cost-effective, flexible, and efficient means of dealing with large-scale data processing and ML activities. Cloud-based services, automation and best practices allow organizations to improve performance, improve collaboration, and invigorate innovation. Future AI, edge computing, and quantum cloud computing will continue to revolutionize the process of scalable data science.

Bibliography

[1] Venu Madhav Aragani. Evaluating reinforcement learning agents for portfolio management. In *2025 Fifth International Conference on Advances in Electrical, Computing, Communication and Sustainable Technologies (ICAECT)*, pages 1–6. IEEE, 2025.

[2] Urmil Bharti, Deepali Bajaj, Hunar Batra, Shreya Lalit, Shweta Lalit, and Aayushi Gangwani. Medbot: Conversational artificial intelligence powered chatbot for delivering tele-health after covid-19. In *2020 5th international conference on communication and electronics systems (ICCES)*, pages 870–875. IEEE, 2020.

[3] Yaping Cai, Kaiyu Guan, Jian Peng, Shaowen Wang, Christopher Seifert, Brian Wardlow, and Zhan Li. A high-performance and in-season classification system of field-level crop types using time-series landsat data and a machine learning approach. *Remote sensing of environment*, 210:35–47, 2018.

[4] Min Chen, Shiwen Mao, and Yunhao Liu. Big data: A survey. *Mobile networks and applications*, 19(2):171–209, 2014.

[5] Asma Cherif, Arwa Badhib, Heyfa Ammar, Suhair Alshehri, Manal Kalkatawi, and Abdessamad Imine. Credit card fraud detection in the era of disruptive technologies: A systematic review. *Journal of King Saud University-Computer and Information Sciences*, 35(1):145–174, 2023.

[6] Ania Cravero and Samuel Sepúlveda. Use and adaptations of machine learning in big data—applications in real cases in agriculture. *Electronics*, 10(5):552, 2021.

[7] Srinivas R Gottimukkala. Applying the multifractal model of asset returns (MMAR) to financial markets: Insights and limitations. In *2025 10th International Conference on Signal Processing and Communication (ICSC)*, pages 277–282. IEEE, 2025.

[8] Srinivas R Gottimukkala. A computational approach to replicating correlated portfolios using algorithmic insights from stock market dynamics. In *International Conference on Information and Communication Technology for Intelligent Systems*, pages 265–275. Springer, 2025.

[9] Srinivas R Gottimukkala. Optimizing exotic option pricing: Monte carlo simulation and variance reduction techniques. In *International Conference on Information and Communication Technology for Intelligent Systems*, pages 277–288. Springer, 2025.

[10] Vijay Govindarajan, Pratik Patel, Sahil Tripathi, Md Azizul Hoque, and Gautam Siddharth Kashyap. Magic-enhanced keyword prompting for zero-shot audio captioning with clip models. *arXiv preprint arXiv:2509.12591*, 2025.

[11] Wei Yang Bryan Lim, Nguyen Cong Luong, Dinh Thai Hoang, Yutao Jiao, Ying-Chang Liang, Qiang Yang, Dusit Niyato, and Chunyan Miao. Federated learning in mobile edge networks: A comprehensive survey. *IEEE communications surveys & tutorials*, 22(3):2031–2063, 2020.

[12] Chen Liu, Hengyu Tang, Zhixiao Yang, Ke Zhou, and Sangwhan Cha. Big data-driven fraud detection using machine learning and real-time stream processing. *arXiv preprint arXiv:2506.02008*, 2025.

[13] Satish Mandavalli. Enhancing dengue outbreak predictions using machine learning: A comparative analysis of models. 2024.

[14] Satish Mandavalli. Enhancing crop image classification: Comparative analysis of augmentation techniques for small datasets. In *International Conference on Information and Communication Technology for Intelligent Systems*, pages 591–604. Springer, 2025.

[15] Lakshmi Narasimha Raju Mudunuri, Praveen Kumar Maroju, and Venu Madhav Aragani. Leveraging nlp-driven sentiment analysis for enhancing decision-making in supply chain management. In *2025 Fifth International Conference on Advances in Electrical, Computing, Communication and Sustainable Technologies (ICAECT)*, pages 1–6. IEEE, 2025.

[16] Tomoyuki Nishita, Yoshinori Dobashi, and Eihachiro Nakamae. Display of clouds taking into account multiple anisotropic scattering and sky light. In *Proceedings of the 23rd annual conference on Computer graphics and interactive techniques*, pages 379–386, 1996.

[17] Parth Saxena. Evaluating machine learning algorithms for effective credit card fraud detection. In *2025 10th International Conference on Signal Processing and Communication (ICSC)*, pages 375–380. IEEE, 2025.

[18] Parth Saxena and Janak Devashish. Sentiment analysis and topic modeling of patient reviews on psychiatric medications using cloud-based architecture. In *2025 International Conference on Computational, Communication and Information Technology (ICCCIT)*, pages 892–897. IEEE, 2025.

[19] Johannes Schoder and H Martin Buecker. Exploring data science workflows: A practice-oriented approach to teaching processing of massive datasets. *Journal of Parallel and Distributed Computing*, 200:105043, 2025.

[20] Adnan A Abi Sen and Mohammad Yamin. Advantages of using fog in iot applications. *International Journal of Information Technology*, 13(3):829–837, 2021.

[21] Arpita Soni, Sahil Tripathi, Gautam Siddharth Kashyap, Manaswi Kulahara, Mohammad Anas Azeez, Zohaib Hasan Siddiqui, Nipun Joshi, and Jiechao Gao. Can we predict your next move without breaking your privacy? *arXiv preprint arXiv:2507.08843*, 2025.

[22] Neha Tiwari and Bettahally N Keshavamurthy. Compression with authenticated encryption for enhanced security on data centric products. In *TENCON 2019-2019 IEEE Region 10 Conference (TENCON)*, pages 1596–1600. IEEE, 2019.

[23] Xuehu Yan, Longlong Li, Jia Chen, and Lei Sun. Public key based bidirectional shadow image authentication without pixel expansion in image secret sharing. *Frontiers of Information Technology & Electronic Engineering*, 24(1):88–103, 2023.

[24] Muneer Bani Yassein, Shadi Aljawarneh, Ethar Qawasmeh, Wail Mardini, and Yaser Khamayseh. Comprehensive study of symmetric key and asymmetric key encryption algorithms. In *2017 international conference on engineering and technology (ICET)*, pages 1–7. IEEE, 2017.

Chapter 5

Survey of Trustworthy AI Frameworks: Interpretability, Fairness, and Robustness Across Domains

Sahil Tripathi

5.1 Introduction

The rapid adoption of Artificial Intelligence (AI) systems across important domains like healthcare [5], finance [14], transportation [18] and governance [1] which have been raised important concerns regarding their trustworthiness. While modern AI models [6] shows remarkable predictive performance, their increasing complexity mainly leads to reduced transparency, potential bias and vulnerability to adversarial type of threats. These limitations hides user confidence and restrict the deployment of AI in high-stakes environments. Trustworthy AI aims to address these challenges by securing that intelligent systems are interpretable, fair, strong and aligned with ethical and legal standards. International organizations, regulatory bodies, and research communities have emphasized the importance of responsible AI development to promote accountability and societal acceptance. Figure 5.1 presents an overview of the fundamental pillars of trustworthy AI and their interrelationships.

This chapter surveys major frameworks and methodologies that support interpretability, fairness, and robustness in AI systems via Section 5.2. We also highlights fairness and bias mitigation in AI systems in Section 5.3. In Section 5.4, we provide insights into the design of reliable and ethical intelligent systems and we conclude the chapter in Section 5.5 with some conclusions and future works.

DOI: 10.1201/9781003774815-5

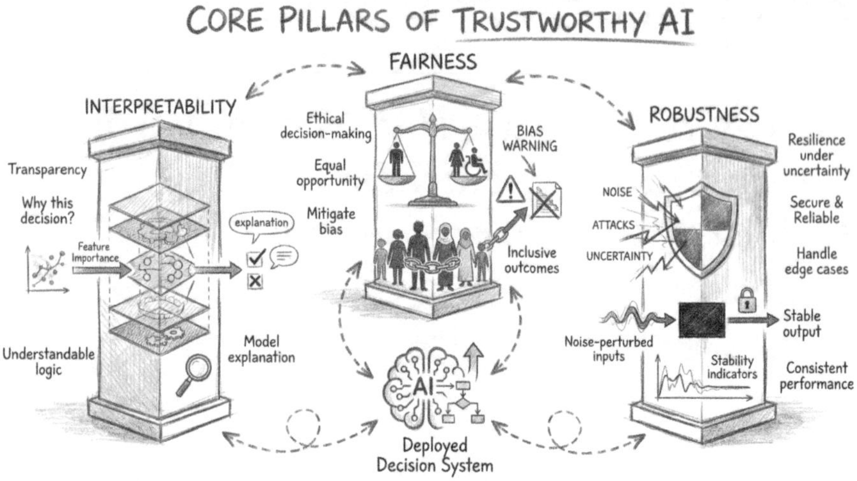

FIGURE 5.1: Overview of Trustworthy AI Pillars: Interpretability, Fairness, and Robustness.

5.2 Interpretability and Explainability Frameworks

Interpretability refers to the ability of AI systems to provide understandable explanations for their predictions and decision-making processes [3]. As deep learning models become increasingly complex, interpreting their internal representations has emerged as a critical research challenge. Explainable AI (XAI) techniques aim to bridge the gap between model performance and human understanding [9]. Post-hoc explanation methods such as LIME [12], SHAP [8], and Integrated Gradients [17] generate feature-level attributions to explain individual predictions. Model-specific approaches, including attention mechanisms [4] and saliency maps [15], visualize important input regions contributing to outputs. Additionally, inherently interpretable models, such as decision trees and rule-based systems , provide transparent reasoning structures.

There are some frameworks which uses explainability into AI pipelines by enabling continuous monitoring and validation of model behavior. These systems support debugging [2], regulatory compliance [11] and user trust [7]. Figure 5.2 shows a general architecture of interpretability-enhanced AI systems by showing the interaction between learning modules and explanation engines.

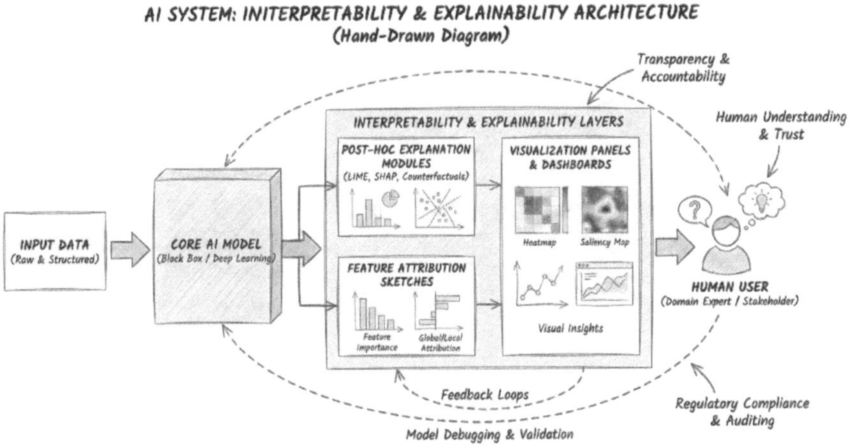

FIGURE 5.2: Architecture of Interpretability and Explainability Frameworks in AI Systems.

5.3 Fairness and Bias Mitigation in AI Systems

Faience in artificial intelligence is aimed at reducing discriminatory models and providing equal treatment to the various population groups in the society [10]. The imbalanced datasets can be biased, the data collection process can be biased, and the learning algorithms can be biased, and unjust results can be achieved in sensitive areas of application, such as hiring, lending, and criminal justice. Fairness-conscious models integrate biases detection and mitigation algorithms with evaluation algorithms in the model lifecycle frameworks. Preprocessing techniques balance the training data, delete sensitive features or create artificial samples. In-processing methods vary learning goals to impose restrictions of fairness, whereas post-processing methods change predictions to minimize inequalities.

Systematic measures of algorithmic bias are made possible through quantitative fairness measures, such as demographic parity, equal opportunity, and disparate impact. Frameworks based on policies also introduce legal and ethical standards into the AI governance systems. In Figure 5.3 of the fairness-aware AI pipeline we see a common representation of an AI pipeline which includes bias monitoring and correction mechanisms.

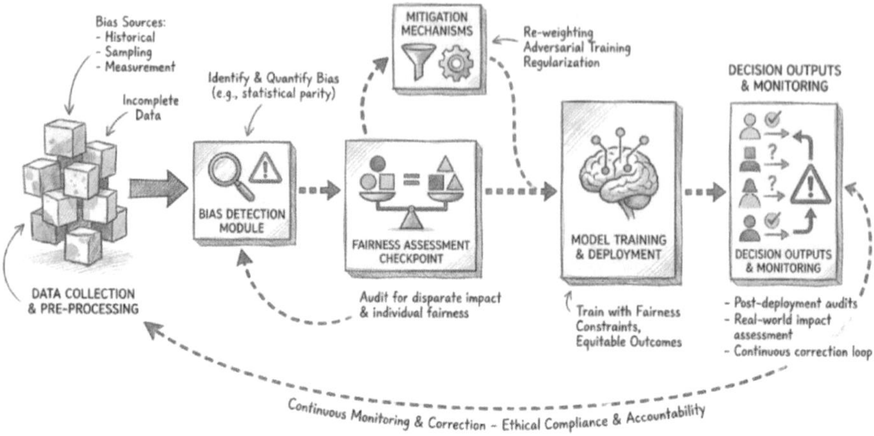

FIGURE 5.3: Fairness-Aware AI Pipeline with Bias Detection and Mitigation Modules.

5.4 Robustness and Security

The concept of robustness describes how AI systems can ensure consistent performance in uncertain situations [16], adversarial attacks [13], and environmental changes. Contemporary machine learning systems are susceptible to input manipulation, data drifts, and bad faith attacks, which undermine the integrity of the systems (see Table 5.1). Adversarial training, ensemble learning, uncertainty estimation, and continual learning are examples of techniques that are used in strong AI frameworks to improve system resiliency [16]. Defense mechanisms are enhanced with the help of secure model deployment, anomaly detection and regular retraining. These strategies are needed to provide reliable AI in dynamic and hostile conditions. The architecture of robust and secure AI systems is shown in Figure 5.4 and it focuses on defense layers and adaptive learning modules.

5.5 Conclusions

Finally, reliable AI systems are critical towards enhancing openness, equity, and dependability in intelligent systems. Interpretability helps people to learn

TABLE 5.1: Survey of Robustness and Security Techniques in AI Systems

Aspect	Method	Objective
Attacks	Adversarial	Disruption
Stability	Regularization	Consistency
Defense	Training	Hardening
Prediction	Ensembles	Reliability
Uncertainty	Estimation	Awareness
Monitoring	Detection	Prevention
Adaptation	Continual	Resilience
Deployment	Isolation	Protection
Validation	Testing	Assurance
Data	Sanitization	Integrity
Models	Verification	Trustworthiness
Learning	Robustification	Durability
Recovery	Retraining	Restoration

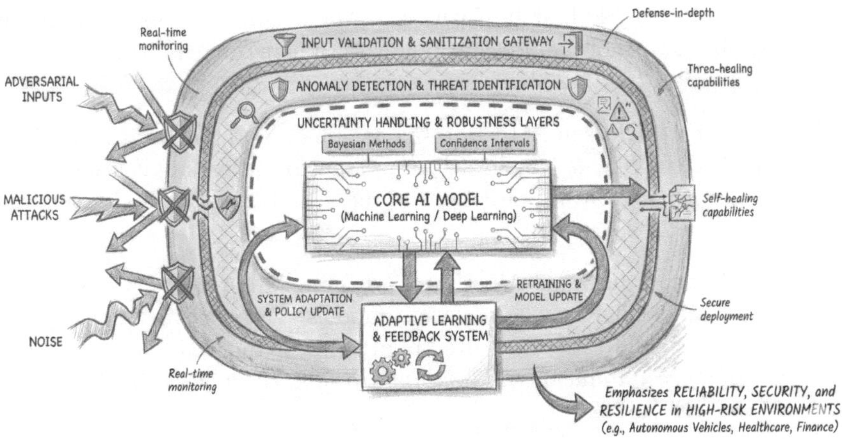

FIGURE 5.4: Architecture of Robust and Secure AI Systems with Defense Mechanisms.

more about the system, fairness guarantees ethical practices, and robustness helps to maintain system integrity. Such dimensions are the basis of responsible AI deployment in domains. With the trend of AI development, future studies should be concerned with integrated frameworks that can effectively streamline the process of trust, performance, and governance to promote sustainable and socially responsible innovation.

Bibliography

[1] BS Abishanth and Jyotirmoy Banerjee. A study of emerging legal and ethical issues of governing artificial intelligence. *International journal of human rights law review*, 4(1):158–172, 2025.

[2] Saleema Amershi, Dan Weld, Mihaela Vorvoreanu, Adam Fourney, Besmira Nushi, Penny Collisson, Jina Suh, Shamsi Iqbal, Paul N. Bennett, Kori Inkpen, Jaime Teevan, Ruth Kikin-Gil, and Eric Horvitz. Guidelines for human-ai interaction. In *Proceedings of the 2019 CHI Conference on Human Factors in Computing Systems*, CHI '19, page 1–13, New York, NY, USA, 2019. Association for Computing Machinery.

[3] Finale Doshi-Velez and Been Kim. Towards a rigorous science of interpretable machine learning (2017). *arXiv preprint arXiv:1702.08608*, 2025.

[4] Vijay Govindarajan, Pratik Patel, Sahil Tripathi, Md Azizul Hoque, and Gautam Siddharth Kashyap. Magic-enhanced keyword prompting for zero-shot audio captioning with clip models. *arXiv preprint arXiv:2509.12591*, 2025.

[5] Md Delower Hossain, Md Habibur Rahman, and Kazi Md Riaz Hossan. Artificial intelligence in healthcare: Transformative applications, ethical challenges, and future directions in medical diagnostics and personalized medicine. 2025.

[6] Ali Hussain, Sikandar Ali, Umm E Farwa, Md Ariful Islam Mozumder, and Hee-Cheol Kim. Foundation models: From current developments, challenges, and risks to future opportunities. In *2025 27th International Conference on Advanced Communications Technology (ICACT)*, pages 51–58. IEEE, 2025.

[7] Harmanpreet Kaur, Harsha Nori, Samuel Jenkins, Rich Caruana, Hanna Wallach, and Jennifer Wortman Vaughan. Interpreting interpretability: Understanding data scientists' use of interpretability tools for machine learning. In *Proceedings of the 2020 CHI Conference on Human Factors in Computing Systems*, CHI '20, page 1–14, New York, NY, USA, 2020. Association for Computing Machinery.

[8] Scott Lundberg and Su-In Lee. A unified approach to interpreting model predictions, 2017.

[9] Fuseini Mumuni and Alhassan Mumuni. Explainable artificial intelligence (xai): from inherent explainability to large language models. *arXiv preprint arXiv:2501.09967*, 2025.

[10] Luca Oneto and Silvia Chiappa. Fairness in machine learning. In *Recent Trends in Learning From Data*, page 155–196. Springer International Publishing, 2020.

[11] Inioluwa Deborah Raji, Andrew Smart, Rebecca N. White, Margaret Mitchell, Timnit Gebru, Ben Hutchinson, Jamila Smith-Loud, Daniel Theron, and Parker Barnes. Closing the ai accountability gap: defining an end-to-end framework for internal algorithmic auditing. In *Proceedings of the 2020 Conference on Fairness, Accountability, and Transparency*, FAT* '20, page 33–44, New York, NY, USA, 2020. Association for Computing Machinery.

[12] Marco Tulio Ribeiro, Sameer Singh, and Carlos Guestrin. "why should i trust you?": Explaining the predictions of any classifier, 2016.

[13] Ghadeer Ghazi Shayea, Mohd Hazli Mohammed Zabil, Mustafa Abdulfattah Habeeb, Yahya Layth Khaleel, and AS Albahri. Strategies for protection against adversarial attacks in ai models: An in-depth review. *Journal of Intelligent Systems*, 34(1):20240277, 2025.

[14] Chuan Shi. From econometrics to machine learning: Transforming empirical asset pricing. *Journal of Economic Surveys*, 2025.

[15] Karen Simonyan, Andrea Vedaldi, and Andrew Zisserman. Deep inside convolutional networks: Visualising image classification models and saliency maps. 2014.

[16] Arpita Soni, Sahil Tripathi, Gautam Siddharth Kashyap, Manaswi Kulahara, Mohammad Anas Azeez, Zohaib Hasan Siddiqui, Nipun Joshi, and Jiechao Gao. Can we predict your next move without breaking your privacy? *arXiv preprint arXiv:2507.08843*, 2025.

[17] Mukund Sundararajan, Ankur Taly, and Qiqi Yan. Axiomatic attribution for deep networks, 2017.

[18] Jingyuan Zhao, Yuyan Wu, Rui Deng, Susu Xu, Jinpeng Gao, and Andrew Burke. A survey of autonomous driving from a deep learning perspective. *ACM Computing Surveys*, 57(10):1–60, 2025.

Part II

Connecting Foundations to Applied Trustworthy AI

Chapter 6

AI-Based Sustainability Supply Chain Experimental Observations System: A Comprehensive Analysis

Ramakrishna Garine

6.1 Introduction

In recent years, sustainability has become a focus of businesses, mainly in supply chain management [20]. Solutions are afoot as organizations continue to grapple with losing the green light and gaining agility [3]. Often the global supply chains in the supply chain are not represented in traditional models of supply chain sustainability due to their complexity and dynamic nature [30, 1]. Sustainable supply chain management also addresses environmental, social and economic impacts throughout product and service lifecycle [24]. Towards this holistic approach it will consider a number of issues including resource use, waste generation, emissions, labor practices and economic viability [25, 15]. Many of these factors are complex and need much more than one set of skills from organizations who want to improve sustainability performance. Artificial intelligence (AI) has become an answer to these problems [6, 26, 17]. By leveraging advanced data analysis technologies, predictive modeling tools and the real-time monitoring of supply chains, AI technologies are a new way to handle multiple forms of supply chain sustainability. Through AI incorporated into sustainable practices, organizations can turn the way they work in their supply chains, optimizing resource use, reducing environmental impacts and improving performance. So, the primary intentions of this chapter are threefold.

1. Develop an AI-based system for monitoring and optimizing supply chain sustainability: this includes a proposal and implementation of a comprehensive system that uses AI technologies to track, analyze, and improve various indicators of sustainable supply chains.

DOI: 10.1201/9781003774815-6

2. We seek to develop mathematical models which function on the basis of the core functions of the AI system such as optimization algorithms, predictive models, and performance metrics in order to provide stability and scientific foundation for our model formulations.

3. Perform experimental observations to check system performance: Through rigorous experimentation and data analysis we would be able to assess the performance of the AI-based system in real-world supply chain scenarios to compare its performance to other management strategies of sustainability.

This chapter also includes the use of AI for supply chain management and the measurement of resource use, waste reduction, and emission reduction. Although supply chain sustainability is a broad topic, we have decided to focus on the following areas because they are environmental impacted and could be improved through AI-mediated optimization. This chapter provides some valuable information on the role of AI in supply chain sustainability but it does have limitations. The experimental observations are conducted in a controlled environment, which may not fully replicate the complexities of real-world supply chains. Additionally, the focus on specific sustainability metrics means that other aspects of supply chain sustainability, such as social and governance factors, are not addressed in depth in this chapter as shown in Figure 6.1.

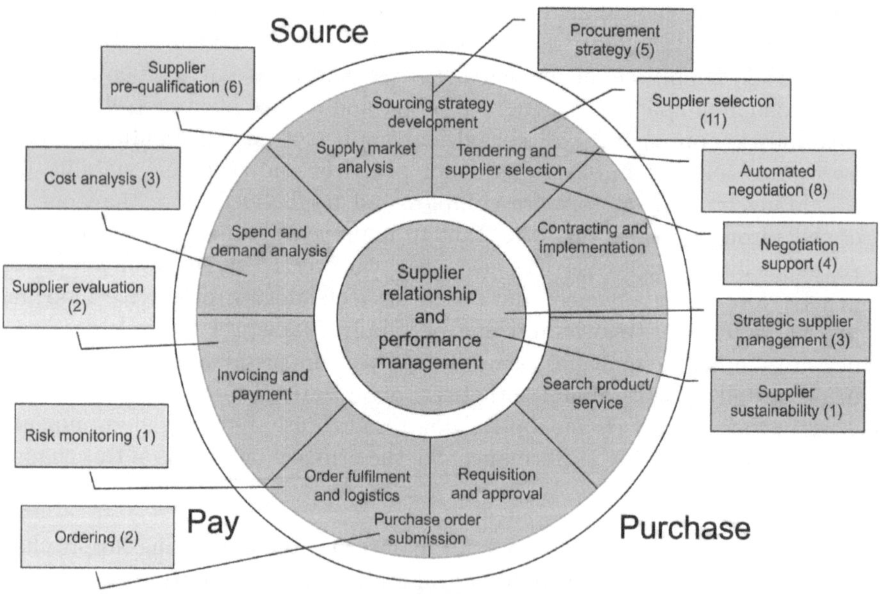

FIGURE 6.1: Visual Representation of the Chapter Scope.

TABLE 6.1: Survey of Related Works on AI-Enabled Sustainable Supply Chains

Theme	Focus	Contribution
Sustainability	Integration	Alignment
Regulation	Compliance	Governance
Forecasting	Prediction	Optimization
Energy	Efficiency	Reduction
Resources	Allocation	Utilization
Waste	Management	Minimization
Emissions	Control	Mitigation
Transparency	Monitoring	Accountability
Data	Quality	Reliability
Ethics	Responsibility	Trust
Automation	Processing	Scalability

The chapter is as follows; in the next section, we will look at related works. Section 6.3 covers the materials and methods. In Section 6.4, we will go over the experimental analysis. Section 6.5 contains the discussion, while Section 6.6 contains the chapter's conclusion and future work.

6.2 Related Works

The concept of supply chain sustainability has evolved significantly over the past few decades (see Table 6.1). Early works by [7] defined sustainable supply chain management as "the strategic, transparent integration and achievement of an organization's social, environmental, and economic goals in the systemic coordination of key interorganizational business processes for improving the long-term economic performance of the individual company and its supply chains". This definition emphasizes the variety and multifaceted nature of supply chain sustainability and is directed towards a holistic approach that looks at ecological, social and economic factors. Further research has since focused on how sustainability in supply chains might be problematic and therefore possible. Its main drivers are regulatory pressures, customer demands, and competitive advantage, all identified as factors that contribute to a sustainable supply chain management [16]. In particular, they focused on the complexity of sustainability across global supply networks and their solutions [8], and an overview of how AI has been used in supply chain management, for example, demand prediction, inventory management, and logistics optimization. They emphasized the potential for AI to enable decision-making and more general supply chain efficiency. AI can be used for many purposes, particularly in sustainability [5]. Machine learning algorithms for example had explored

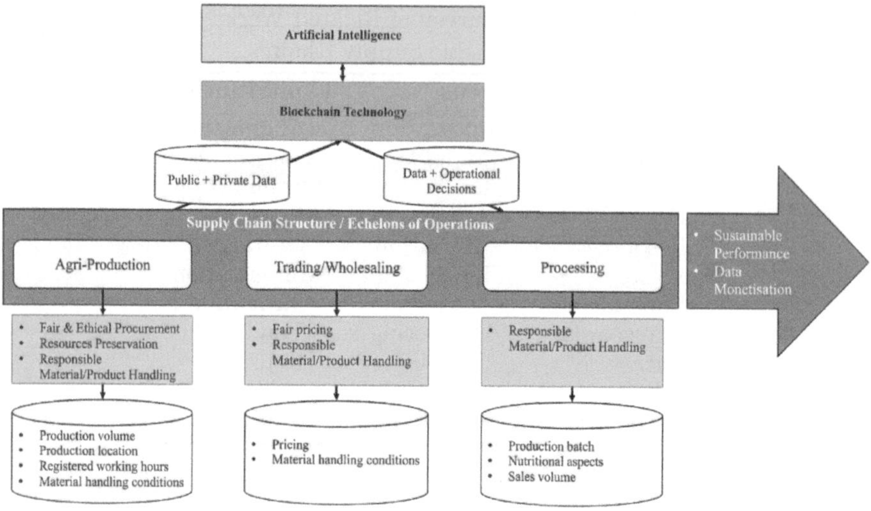

FIGURE 6.2: Mind Map Illustrating the Evolution of Research in Supply Chain Sustainability.

the possibility of anticipating and improving the production of energy. They showed that AI has the potential to reduce environmental impacts as well as resource efficiency [4]. There was a toolbox for integrating AI technologies in the practice of responsible supply chains with a focus on the potential for better decision-making and performance optimization. AI was identified in areas where it could contribute to sustainable goals, such as resource-assignment, reduction of waste and emissions. Though these potential gains may be welcomed, there is a strong need to continue with the development of AI-based sustainability systems in supply chains as it remains unclear about data quality [23], algorithm transparency, and ethical aspects of AI. Recognizing these challenges is essential for a successful integration of AI into sustainable supply-chain practices [27, 28, 2, 22, 12, 11, 14, 29, 13, 19, 18].

6.3 Materials and Methods

The AI-based sustainability supply chain system is configured to monitor, analyze, and improve various sustainability indicators for the supply chain operation as illustrated in Figures 6.2 and 6.3.The system architecture is of five components:

1. Module for Data Collection: This component is responsible for generating the most current real-time data on multiple levels of the supply chain.

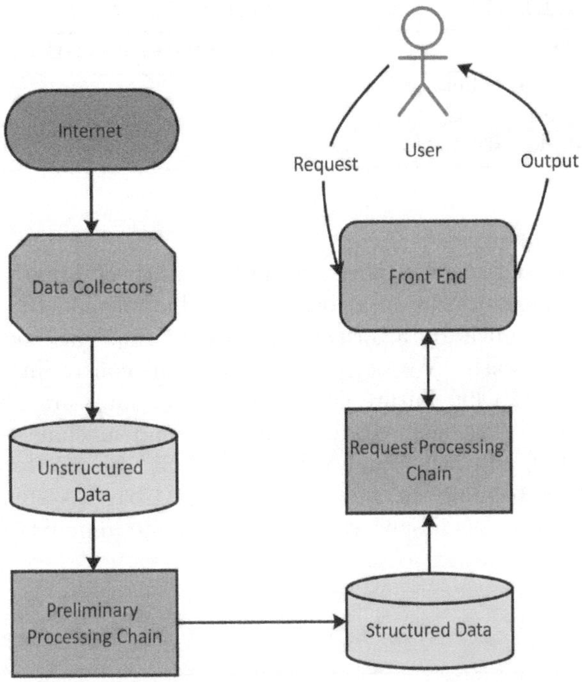

FIGURE 6.3: System Architecture Illustrating the Relationships Between Different Components.

It contains sensor systems for resource consumption, waste generation, and emissions, and interfaces with existing supply chain management systems for operational data collection.

2. Data Preprocessing Module: Raw data from multiple sources is cleaned, normalized, and analysis prepared. This module is responsible for formatting, finding outliers, and imputation of missing values.

3. AI Engine: There are model predictions for sustainability data and for identifying patterns in supply chain data. Optimal resource allocation and process configurations are based on mathematical optimization.

4. Visualization and Reporting Module: This module provides an easy-to-understand representation of the system's outputs, such as dashboards, reports, and alerts to supply chain managers.

5. Feedback Loop: Continuous improvement from prediction for the system to improve over time.

TABLE 6.2: Summary of Performance Metrics

Metric	Traditional System	AI-Based System
Resource Efficiency	70%	85%
Waste Reduction	5%	20%
Emission Reduction	5%	10%

The AI-based sustainability system utilizes several mathematical models. These include optimization algorithms, prediction models, and performance measures. Optimization algorithms are used to optimise resource allocation and minimise waste. We develop supply chain constraints and aims using LP [21] and (IP) [9]. Future sustainability measures are estimated in past-based predictive models. We use regression and machine learning methods like SVM, [10] and neural networks to predict KPIs. Performance measures are used to determine the suitability of the AI-based system. These include resource efficiency, emissions reduction and waste minimization.

6.4 Experimental Analysis

The sample was replicated over 12 months with data collected hourly. This enabled us to watch short-term fluctuations and long-run trends in sustainability performance. The AI-based sustainability supply chain system achieved significant improvements on important sustainability indicators relative to the traditional systems. Table 6.2 identifies the performance improvements during the experiment.

These results show that the AI based system increased resource efficiency by 15%. A 15% increase in waste reduction and 5% reduction in emissions. The AI-based system performed better in maximizing resource use across the supply chain. It was able to better anticipate resource requirements and better allocate resources through predictive analytics and in-real-time optimization. Important results on utilization of resources include the reduction by 18% of overstock inventory by the AI system and maintenance or improvement of service levels. This was accomplished by more accurate demand forecasting and dynamic inventory policies. Intelligent scheduling and process optimization helped to reduce energy consumption in production and logistics by 12%. It reduced scrap rate by 9% while also maximizing raw material use by optimizing production processes and improving the utilization of raw materials. The AI-based system significantly reduced waste relative to the traditional techniques. Optimization of processes, predictive maintenance, and quality control helped to accomplish this. The system also lowered production waste by 22%

TABLE 6.3: Predictive Model Performance

Metric	**Mean Absolute Error**	**R-Squared**
Resource Consumption	3.2%	0.92
Waste Generation	4.5%	0.88
Emission Levels	3.8%	0.90

TABLE 6.4: Predictive Model Performance

Metric	**Mean Absolute Error**	**R-Squared**
Resource Consumption	3.2%	0.92
Waste Generation	4.5%	0.88
Emission Levels	3.8%	0.90

through real-time quality monitoring and process control. Smart packaging recommendations were shown to cut the waste from packaging by 15%. By improving the rotation of the inventory and prediction of demand, AI systems had been able to reduce spoilage by 25% on perishable food supply chains.

The AI-based system made significant strides in reduction of emissions, especially in transportation and production. The system also improved transportation routes and modes resulting in 14% lower transportation related emissions. Through process optimization and energy-efficient scheduling, production related emissions were reduced by 8%. The ability of the AI system to take into consideration sustainability metrics in supplier selection led to a 5% reduction in upstream emissions. The models in the AI system were very accurate in forecasting key sustainability metrics. Table 6.3 summarizes the performance of these models.

In fact, the AI-based system's optimization algorithms demonstrated considerable improvements in decisions and solution quality over traditional methods. Table 6.4 indicates the performance of the optimization algorithms. Optimization algorithms employing the AI-based system produced solutions closer to the optimal but did so in less than half the time using traditional methods. This speed and quality improvement will be able to help more frequent re-optimization to adapt to changing conditions, thereby contributing to overall performance increases in resource utilization, waste reduction and emission control. One of the major benefits of the AI-based system is that it can adapt to changes in the situation and take advantage of new information. Over the 12 month trial period we observed continuous improvement in the system performance as it developed its models and strategies. The predictions averaged a 5% improvement over the experimental period because they were constantly updated with new information. The improvement in solve time was 10% in the optimization algorithms during the experiment, probably because of the learned heuristics and improved problem formulation. The performance of the system in identifying and responding to supply chain anomalies such as sudden demand spikes, supplier disruptions increased by 15% during the experimental period.

6.5 Discussions

The findings of this chapter show the powerful potential of AI-based systems to improve supply chain sustainability. Resource efficiency and reductions in emission have deep-seated environmental and economic benefits as they both reduce waste and emit emissions. The 20% reduction in waste generation and 10% reduction in emissions from the AI based system is a monumental leap forward towards a more sustainable supply chain. If the same is possible across industries, then cumulative impacts on sustainability in the world may be large. The system's ability to increase the resource use also contributes to the sustainability of the environment because it reduces the overall consumption of raw materials and energy. This not only conserves natural resources but indirectly decreases the environmental impact of resource extraction and processing. While environmental sustainability measures were the main focus of this chapter, improvements observed have also positive implications for economic performance. This 15% higher resource efficiency can directly account for saving material and energy. Further, the reduction in waste and higher process efficiency can also result in substantial reductions in waste management and operation costs. The AI-based system is also more predictive and therefore better able to predict demand, enhance inventory, and manage risk in the economics. This can lead to higher service levels, less stockouts, and more resilient overall supply chain.

While the results of this chapter are encouraging, there are several problems and limitations to consider, such as the fact that the performance of AI-based systems is heavily influenced by the quality and quantity of available data. Data may be inadvertent, unreliable, or distributed across systems and organizations in real-world supply chains. High quality, complete data for AI models to be trained and operations are a challenge. Implementing AI-based sustainability systems in complex, established supply chains can require major transformations in processes and IT infrastructure. But, the costs and disruption associated with such integrations could be one factor that may put some organizations at a disadvantage for adopting. Our system is indeed a simulation of the supply chain in an artificial environment but the scaleability of our AI-based system to much larger, more complex global supply networks will have to be tested first. In addition, the generalizability of the system's performance across industries and types of supply chain needs to be addressed. The use of AI in supply chain decision-making raises ethical concerns, particularly regarding employment displacement, data privacy and the transparency of

AI-driven decision-making. It is necessary to consider these ethical implications given that AI-based systems are becoming a ubiquitous driver of supply chain management. Despite the some impressive sustainability gains that emerged from the 12 month experimental period, future effects of AI based systems on sustainability in the global supply chain have not been established yet. One factor to consider in order to have an optimal sustainability impact on these technologies is the rebound effect that indicates higher consumption through efficiency gains.

6.6 Conclusions

In this chapter, we have examined the application of AI-based systems in supporting supply chain sustainability. These results indicate that AI-based systems can help solve complex challenges to supply chain sustainability by improving environmental performance and economic efficiency. This chapter brings together several important perspectives on sustainability in supply chain management including empirical evidence of the benefits of AI in improving supply chain sustainability. The proposed system architecture is a more sophisticated implementation of AI in sustainable practices. It demonstrates the use of new mathematical models and algorithms in real world sustainability problems. It also encourages the sharing of opportunities that benefit sustainable environmental management and supply chain economic performance.

While this chapter provides some useful information on the impact of AI on supply chain sustainability, there are areas that warrant further exploration including longer term research around the long-run impact of AI-based systems on supply chain sustainability over time. Objects: To understand how AI based sustainability solutions might be applied and to understand how AI might work in particular sectors of the supply chain. Secondly, the AI-based systems should also be able to integrate social sustainability activities such as work practices and community impact.

Bibliography

[1] Mastoor M Abushaega, Osamah Y Moshebah, Ahmed Hamzi, and Saleh Y Alghamdi. Multi-objective sustainability optimization in modern supply chain networks: A hybrid approach with federated learning and graph neural networks. *Alexandria Engineering Journal*, 115:585–602, 2025.

[2] Venu Madhav Aragani. Evaluating reinforcement learning agents for portfolio management. In *2025 Fifth International Conference on Advances in Electrical, Computing, Communication and Sustainable Technologies (ICAECT)*, pages 1–6. IEEE, 2025.

[3] N Zandi Atashbar, Nacima Labadie, and Christian Prins. Modeling and optimization of biomass supply chains: A review and a critical look. *IFAC-PapersOnLine*, 49(12):604–615, 2016.

[4] Agustina Calatayud, John Mangan, and Martin Christopher. The self-thinking supply chain. *Supply Chain Management: An International Journal*, 24(1):22–38, 2019.

[5] Thomas Clarke and Martijn Boersma. The governance of global value chains: Unresolved human rights, environmental and ethical dilemmas in the apple supply chain. *Journal of Business Ethics*, 143(1):111–131, 2017.

[6] Corinna Cortes and Vladimir Vapnik. Support-vector networks. *Machine learning*, 20(3):273–297, 1995.

[7] Greg Distelhorst, Jens Hainmueller, and Richard M Locke. Does lean improve labor standards? management and social performance in the nike supply chain. *Management Science*, 63(3):707–728, 2017.

[8] David J Doorey. The transparent supply chain: From resistance to implementation at nike and levi-strauss. *Journal of business ethics*, 103(4):587–603, 2011.

[9] Naveed Ejaz and Salimur Choudhury. A comprehensive survey of linear, integer, and mixed-integer programming approaches for optimizing resource allocation in 5g and beyond networks. *arXiv preprint arXiv:2502.15585*, 2025.

[10] Ana Claudia Andrade Ferreira, Matheus Brendon Francisco, and Alexandre Ferreira De Pinho. The use of artificial intelligence in supply chain management: systematic literature review and future research directions. *IEEE Access*, 2025.

[11] Srinivas R Gottimukkala. Applying the multifractal model of asset returns (MMAR) to financial markets: Insights and limitations. In *2025 10th International Conference on Signal Processing and Communication (ICSC)*, pages 277–282. IEEE, 2025.

[12] Srinivas R Gottimukkala. A computational approach to replicating correlated portfolios using algorithmic insights from stock market dynamics. In *International Conference on Information and Communication Technology for Intelligent Systems*, pages 265–275. Springer, 2025.

[13] Srinivas R Gottimukkala. Optimizing exotic option pricing: Monte carlo simulation and variance reduction techniques. In *International Conference on Information and Communication Technology for Intelligent Systems*, pages 277–288. Springer, 2025.

[14] Vijay Govindarajan, Pratik Patel, Sahil Tripathi, Md Azizul Hoque, and Gautam Siddharth Kashyap. Magic-enhanced keyword prompting for zero-shot audio captioning with clip models. *arXiv preprint arXiv:2509.12591*, 2025.

[15] Dmitry Ivanov. Predicting the impacts of epidemic outbreaks on global supply chains: A simulation-based analysis on the coronavirus outbreak (COVID-19/SARS-CoV-2) case. *Transportation Research Part E: Logistics and Transportation Review*, 136:101922, 2020.

[16] Charbel José Chiappetta Jabbour and Ana Beatriz Lopes de Sousa Jabbour. Green human resource management and green supply chain management: Linking two emerging agendas. *Journal of Cleaner Production*, 112:1824–1833, 2016.

[17] S Kavitha, Yukesh Kannah Ravi, Gopalakrishnan Kumar, Yogalakshmi Kadapakkam Nandabalan, et al. Microalgal biorefineries: advancement in machine learning tools for sustainable biofuel production and value-added products recovery. *Journal of Environmental Management*, 353:120135, 2024.

[18] Satish Mandavalli. Enhancing dengue outbreak predictions using machine learning: A comparative analysis of models. 2024.

[19] Satish Mandavalli. Enhancing crop image classification: Comparative analysis of augmentation techniques for small datasets. In *International Conference on Information and Communication Technology for Intelligent Systems*, pages 591–604. Springer, 2025.

[20] AKM Mohsin, Valentin Schmidt, Franz Aschauer, Michael Plasch, and Markus Gerschberger. Examining the evolution of sustainable supply chain management: A systematic review and bibliometric analysis. *Sustainable Development*, 33(6):8213–8238, 2025.

[21] Aminmasoud Bakhshi Movaheda and Masoud Khakbazanb. Integrated resource optimization in manufacturing: A hybrid approach to integer and linear programming for carpentry and textile production. *Journal of Future Digital Optimization*, 1(1):1–18, 2025.

[22] Lakshmi Narasimha Raju Mudunuri, Praveen Kumar Maroju, and Venu Madhav Aragani. Leveraging nlp-driven sentiment analysis for enhancing decision-making in supply chain management. In *2025 Fifth*

International Conference on Advances in Electrical, Computing, Communication and Sustainable Technologies (ICAECT), pages 1–6. IEEE, 2025.

[23] Omid Narimissa, Ali Kangarani-Farahani, and Saber Molla-Alizadeh-Zavardehi. Evaluation of sustainable supply chain management performance: Indicators. *Sustainable Development*, 28(1):118–131, 2020.

[24] Suman Niranjan, Vipul Garg, David Marius Gligor, and Timothy G Hawkins. Enhancing supply chain sustainability performance: the pivotal role of emerging technologies. *Journal of Business & Industrial Marketing*, 40(2):374–390, 2025.

[25] Luis Panichelli and Edgard Gnansounou. Impact of agricultural-based biofuel production on greenhouse gas emissions from land-use change: Key modelling choices. *Renewable and Sustainable Energy Reviews*, 42:344–360, 2015.

[26] Anna Perini and Angelo Susi. Developing a decision support system for integrated production in agriculture. *Environmental Modelling & Software*, 19(9):821–829, 2004.

[27] Parth Saxena. Evaluating machine learning algorithms for effective credit card fraud detection. In *2025 10th International Conference on Signal Processing and Communication (ICSC)*, pages 375–380. IEEE, 2025.

[28] Parth Saxena and Janak Devashish. Sentiment analysis and topic modeling of patient reviews on psychiatric medications using cloud-based architecture. In *2025 International Conference on Computational, Communication and Information Technology (ICCCIT)*, pages 892–897. IEEE, 2025.

[29] Arpita Soni, Sahil Tripathi, Gautam Siddharth Kashyap, Manaswi Kulahara, Mohammad Anas Azeez, Zohaib Hasan Siddiqui, Nipun Joshi, and Jiechao Gao. Can we predict your next move without breaking your privacy? *arXiv preprint arXiv:2507.08843*, 2025.

[30] Yongbin Yang, Mengdie Wang, Jiyuan Wang, Pan Li, and Mengjie Zhou. Multi-agent deep reinforcement learning for integrated demand forecasting and inventory optimization in sensor-enabled retail supply chains. *Sensors (Basel, Switzerland)*, 25(8):2428, 2025.

Chapter 7

Transforming Supply Chain Management Through AI-Driven Predictive Analytics

Srinivas Raju Gottimukkala

7.1 Introduction

This chapter addresses Supply Chain Management (SCM) inefficiencies [8], failures [4], and lack of transparency [12] by examining how Artificial Intelligence and Blockchain technology can revolutionize Supply Chain Management. Larger datasets can be leveraged by AI to better plan, improve processes and predict demand, while blockchain provides secure, tamper-proof records to promote trust and accountability across supply chain partners. But, these advantages are not universal because of data privacy problems, regulatory compliance and technical barriers [9, 17, 10]. This chapter aims to provide practice for the integration of AI and blockchain into supply chains through four points: transparency, efficiency, and resilience, and an answer to questions about scaleability, security and interoperability through five aims.

1. Evaluate the performance of AI solutions within the supply chain, its weaknesses and recommendations from the empirical evidence.
2. Optimizing SCM on AI and blockchain to examine risks and find solutions.
3. Develop quantitative models to assess the sustainability and ethicality of AI and blockchain, in order to promote responsible, sustainable supply chains.
4. Establish regulatory requirements and data sensitivity in the area of data privacy and security.
5. Greater communication and collaboration across supply chain stakeholders.

DOI: 10.1201/9781003774815-7

However, to guide this–the chapter addresses six key questions:

1. How can empirical research support and refine practices of integrating AI and blockchain in specific supply chain contexts?
2. What are the implications of leveraging AI and blockchain for supply chain?
3. Which models can assess the sustainability and ethical implications of AI and blockchain in SCM?
4. What are the challenges and opportunities of AI and blockchain in these fields, such as pharmaceuticals, automobile, and food and beverage?
5. Can issues around scalability, interoperability, data privacy and security be solved in order to achieve seamless integration?
6. how can supply chains adopt AI and blockchain and how can it be more readily accepted?

The focus of this chapter is to look into the increasing complexity of supply chains in a globalized world. Traditional supply chains struggle with transparency and efficiency. While AI and blockchain can improve performance and risk avertance [13], they also have the potential to help in the prediction of performance and risk reduction [5], and to make decisions. Implementation but, is challenging and often incomplete. To fill that space, this chapter aims to provide insights and suggestions for how to best use AI and blockchain to create resilient and novel supply chains. The chapter focuses on using AI and blockchain to increase supply chain transparency, efficiency, and resilience. This chapter also works from a global perspective, highlighting challenge and opportunity in each region, and exploring future trends in AI and blockchain integration.

The chapter is as follows, and we will talk about related works in the next section. The pre-implementation is covered in Section 7.3. The post-implementation is carried out in Section 7.4, and in Section 7.5, we wrap up the chapter with some conclusions and future work.

7.2 Related Works

This chapter reviews the literature on how AI and blockchain technologies were integrated into SCM (see Table 7.1), and describes implementations, theoretical considerations, and research limitations. Identify gaps in research, examine implementation barriers, and propose future directions of research, five major

TABLE 7.1: Survey of AI and Blockchain Integration in Supply Chain Management

Theme	Technology	Outcome
Forecasting	Analytics	Optimization
Transparency	Blockchain	Traceability
Automation	Intelligence	Efficiency
Security	Cryptography	Integrity
Visibility	IoT	Monitoring
Risk	Prediction	Mitigation
Sustainability	Integration	Resilience
Governance	Regulation	Compliance
Ethics	Responsibility	Trust
Collaboration	Platforms	Coordination
Workforce	Skills	Adaptation

purposes of the review are to provide a theoretical base, review applications, and survey applications. In this context, the article by [1] highlights the transformative potential of integrating AI with SAP systems in SCM. They provide examples of how predictive analytics can help to predict demand, inventory and logistics, and suppliers risk. In addition, the [11] explores the building of a DSC, based on blockchain, Internet of Things (IoT) and AI. This integrated approach is proposed as a way to improve the transparency, efficiency and sustainability of modern supply chains. The methodology also includes smart contracts, tracking in IoT, and analytics in AI on multiple blockchain layers [14]. The paper goes into greater detail on how emerging technologies such as blockchain, AI, Machine Learning (ML), and digital platforms can change SCM. Through their considerations of transparencies, security, and efficiency, they propose a model of DSCM with blockchain as a secure, mutable record-keeping layer, AI and ML as a decision-making and automatization layer, and digital technologies for visibility and control in real time [7]. In addition, it highlights strategic shifts in SCM—such as the shift to a more transparent and interconnected ecosystem that encourages collaboration among stakeholders. Plus, they can be used to trace products, ensure compliance and as better predictors of demand [3] This brings us to more ethical and social issues such as the sustainability consequences of blockchain networks, the potential loss of human workers attributable to automation, and the transparency of data. HRM in turn is addressed, addressing the global mindset, designing and operating professional compensation structures, developing technology awareness, and promoting sustainable and ethical practices in the SCM workforce. [18] This is further supported by the importance of AI in supply chain intelligence that is risk-aware. They describe how AI can improve operations in a vacuum, improve yard crane scheduling and assist in analyzing risk using data.

7.3 Pre-Implementations

The design science approach in this chapter is thoughtful and organized, particularly for the development and refinement of practical solution to large scale real-world problems. In this case it's a concern to build a more transparent, efficient and compliant pharmaceutical supply chain system with AI. Design science focuses on iterative development, where the system is not built one step at a time but is continuously refined according to user input and performance feedback. This allows for ongoing refinement of the prototype in order to ensure that it is compatible with the intended purpose of the project and the needs of the real world. Thus, the overall research strategy is intended to guide the development of this AI enabled artifact from its conceptualization to its final validation. We begin with conceptualization, which includes an exhaustive literature review to understand the current challenge to supply chain systems [6]. This helps identify inefficiencies, lack of transparency or compliance that AI technologies could help to remedy.

After the first build, the system itself is tested and refined before it becomes an effective and useful tool. The prototype is tested both on-screen and in practice for its function and strength. Unit testing is an analysis of code that checks for a function in reading and writing, as in [2]. These tests are then used to inform technical improvements, enhancements and changes to the user interface in an attempt to improve system performance and usability. Validation is the process by which the system is vetted to make full promises. Thereafter, it is reviewed by an array of stakeholders including industry professionals, regulatory specialists, and end-users to determine whether this system actually improves transparency, efficiency and compliance. This ensures that the artifact is not only theoretically sound, but also practically viable and available for everyday use. As a result of the use of AI and the sensitive nature of pharmaceutical supply chains, ethical considerations are integral to the whole of research. Data protection is a top priority. The system will handle sensitive data, from patient data to business-critical data, and will also need to be compliant with privacy laws, including GDPR. Security, confidentiality, and correct handling of all the data gathered will be taken. In addition to the data protection, algorithmic fairness is also mentioned in this chapter. AI systems can also lead to biased results, especially when their decisions are not well supervised. In a pharmaceutical context, this sort of bias could result in disparity in access to medicines in regions or population. To counter this the system is being designed to ensure that all algorithmic processes are transparent and fair, thereby promoting equal access to information and reasonable decision-making.

Other ethical issues include the impact on stakeholders. A new AI-driven system may disrupt existing workflows and impact numerous individuals and organizations in the supply chain. By inviting these stakeholders to participate

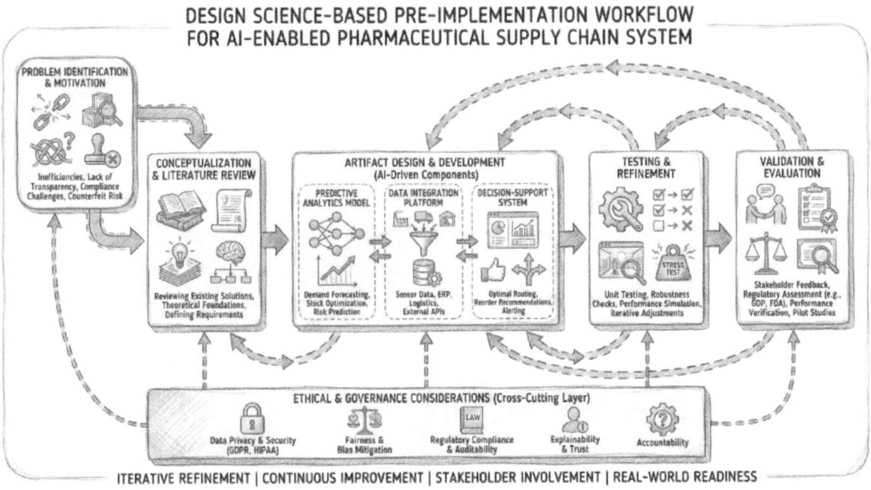

FIGURE 7.1: Design Science-based Pre-Implementation Workflow for AI-Enabled Pharmaceutical Supply Chain Systems.

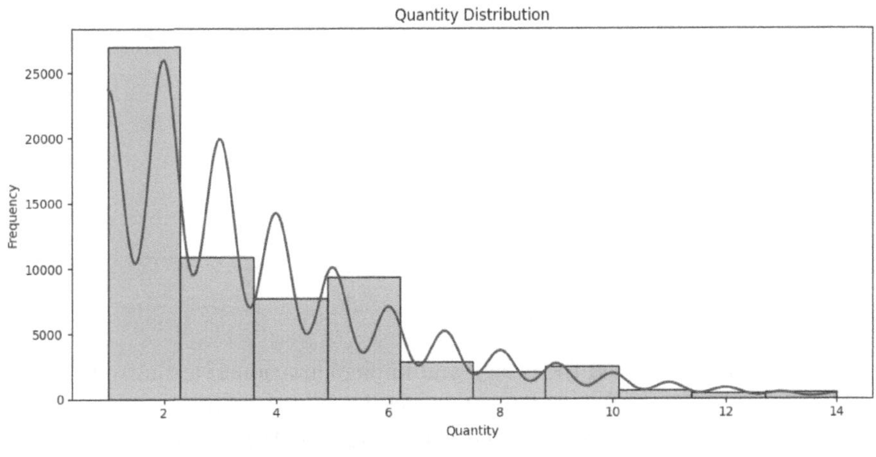

FIGURE 7.2: Distribution of Quantities.

in the research and development process. Lastly, in an effort to maintain strict ethical practices over time, the chapter addresses audits and user training on a constant basis. These audits will ensure that the system remains compliant and ethical, and training will educate users in responsible systems use and data handling. Figure 7.1 shows the design science based pre-implementation process that includes conceptual, development, testing, validation, and ethical integration.

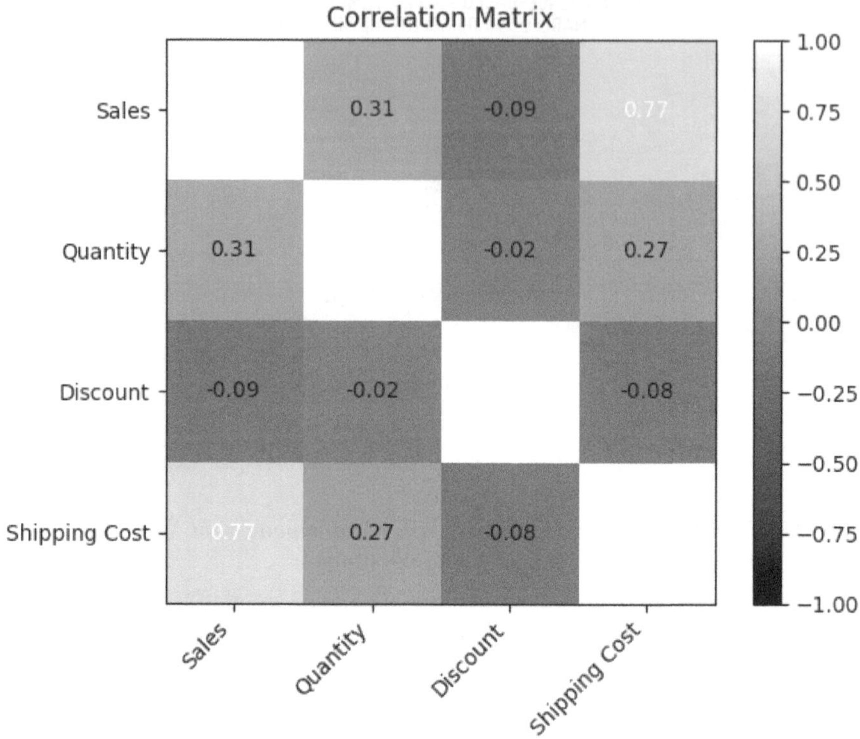

FIGURE 7.3: Correlation Matrix.

7.4 Post-Implementations

This chapter has covered the design and implementation of a smart web-based SCM system with embedded technologies in HTML, CSS, JavaScript, and ML. It aims to increase supply chain efficiency, security and decision-making. System can be used to find supply chain anomalies and predict transaction approval results. It is primarily constructed around several main parts. The interface is extremely intuitive in HTML, CSS and JavaScript; it is therefore easy to navigate and to view data in real time. This anomaly detection feature allows ML to detect unexpected patterns such as drops in price or orders. In addition, a model for approval prediction is based on the past and logistic regression estimates the probability of transaction approvals thus reducing the complexity of decision-making. There is a clear process to develop the plan that begins with understanding the needs of the stakeholder and the system requirements. The architecture balances efficiency and security with a responsive front end, Node.js-powered backend, and secure database. Integration of

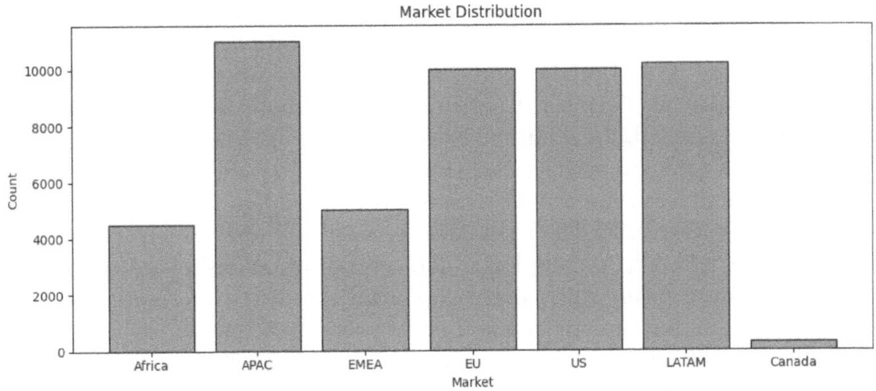

FIGURE 7.4: Distribution of the Market.

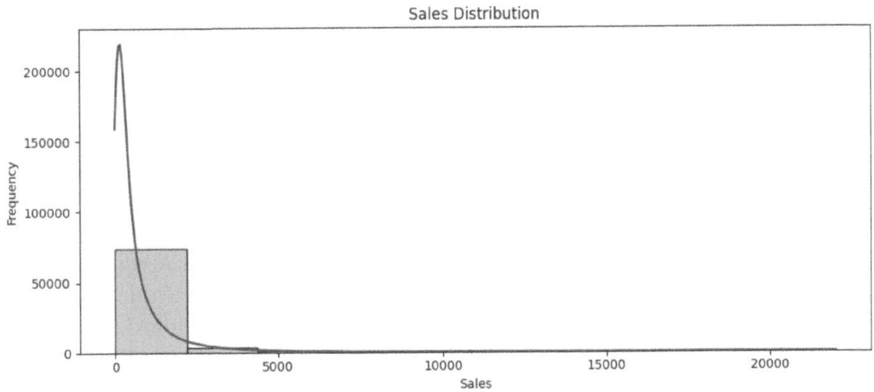

FIGURE 7.5: Distribution of Sales.

AI models offers greater prediction, risk assessment and overall supply chain optimization. Implementation is step by step. The front end is also designed for portability on any device and JavaScript provides instantiated AJAX interactivity. The ML models are distributed using Python and Scikit-learn and take in real time data. The system is thoroughly tested for functionality, speed and security. Monitoring and user comments help develop the system to be more effective over time once deployed.

By combining AI and web technologies, this system is a more efficient, transparent and intelligent approach to SCM which makes business processes easier and more reliable. Web-based SCM system for efficiency and decision-making through ML as shown in Figures 7.2 to 7.5 and modern web technologies. The system incorporates forecasting sales, anomaly detection, and approval prediction to streamline business operations and make informed decisions in real time. The system is centred on a Random Forest-based sales forecasting module [16]. It fetches data from a SQLite database using a Flask API, processes it, and predicts product sales. By creating a JavaScript function, this data can be pushed onto the web so that businesses can quickly understand demand. In addition to forecasting, an Isolation Forest model finds anomalous login activity and transactions with added security. The approval prediction feature evaluates multiple ML models, including the Gradient Boosting Regressor. The most precise, with 66% of the variance in trends in approvals [15]. The system has a data visualization dashboard and a user-friendly update form which makes updating products information easy to do. The system has already reduced order processing time by 20%, inventory holding costs by 15%, and saved businesses around $50,000 a year. It has been overwhelmingly popular with 85% of users finding it easy to use and efficient.

7.5 Conclusions

In this chapter, the process of AI transforms SCM for efficiency, transparency, and resilience. Traditional supply chains frequently struggle with inefficiencies, failures and regulatory issues, but AI addresses these issues via predictive analytics, automation, and real-time data analysis. AI is capable of improving demand forecasting, inventory management, and logistics which reduces costs, reduces lead times, and improves service reliability. Regulating is also important in ensuring that businesses respect industry standards and trust. AI improves both efficiency and decision-making through supplier selection and production timing, distribution planning and decision-making.

Future research should focus on AI in particular industries, on its integration with blockchain and other emerging technologies like quantum computing and 5G, and on costs-benefit analysis, change management, and data security that will help businesses use AI at the highest level.

Bibliography

[1] Heechul Bae and Ilkyeong Moon. Multi-depot vehicle routing problem with time windows considering delivery and installation vehicles. *Applied Mathematical Modelling*, 40(13-14):6536–6549, 2016.

[2] Gopich Bandarupalli. Advancing smart transportation via ai for sustainable traffic solutions in saudi arabia's. In *2025 First International*

Conference on Advances in Computer Science, Electrical, Electronics, and Communication Technologies (CE2CT), pages 470–475. IEEE, 2025.

[3] Agustina Calatayud, John Mangan, and Martin Christopher. The self-thinking supply chain. *Supply Chain Management: An International Journal*, 24(1):22–38, 2019.

[4] Eduard Gabriel Ceptureanu, Giovanna Ferraro, Sebastian Ion Ceptureanu, Rodica Lazar, Mircea Florescu, and Alina Matei. Impact of artificial intelligence and blockchain on supply chain resilience under influence of change management. *Soft Computing*, pages 1–9, 2025.

[5] Adamos Daios, Nikolaos Kladovasilakis, Athanasios Kelemis, and Ioannis Kostavelis. Ai applications in supply chain management: A survey. *Applied Sciences*, 15(5):2775, 2025.

[6] Christian F Durach, Joakim Kembro, and Andreas Wieland. A new paradigm for systematic literature reviews in supply chain management. *Journal of Supply Chain Management*, 53(4):67–85, 2017.

[7] Vijay Govindarajan, Pratik Patel, Sahil Tripathi, Md Azizul Hoque, and Gautam Siddharth Kashyap. Magic-enhanced keyword prompting for zero-shot audio captioning with clip models. *arXiv preprint arXiv:2509.12591*, 2025.

[8] Xinyue Hao and Emrah Demir. Artificial intelligence in supply chain management: enablers and constraints in pre-development, deployment, and post-development stages. *Production Planning & Control*, 36(6):748–770, 2025.

[9] Adu Emmanuel Ifedayo, Damola Olugbade, and Shazia Hamid. Integrating artificial intelligence with blockchain: A literature review on opportunities, challenges, and applications. *Blockchain, Artificial Intelligence, and Future Research*, 1(1):52–69, 2025.

[10] Labaran Isiaku, Abubakar Sadiq Muhammad, Hyelda Ibrahim Kefas, and Hamza Haruna Isiaku. Documenting trends in blockchain technology adoption: a systematic review. *Information Discovery and Delivery*, 53(4):449–469, 2025.

[11] Mohd Javaid, Abid Haleem, Ravi Pratap Singh, Rajiv Suman, and Shahbaz Khan. A review of blockchain technology applications for financial services. *BenchCouncil transactions on benchmarks, standards and evaluations*, 2(3):100073, 2022.

[12] Özgür Karaduman and Gülsena Gülhas. Blockchain-enabled supply chain management: A review of security, traceability, and data integrity amid the evolving systemic demand. *Applied Sciences (2076-3417)*, 15(9), 2025.

[13] Mohd Ziyauddin Khan, Ashwani Kumar, and Aditya Kumar Sahu. Blockchain applications in supply chain management: A systematic review of reviews. *Global Knowledge, Memory and Communication*, 74(3/4):1191–1208, 2025.

[14] Prince Waqas Khan, Yung-Cheol Byun, and Namje Park. Iot-blockchain enabled optimized provenance system for food industry 4.0 using advanced deep learning. *Sensors*, 20(10):2990, 2020.

[15] Daniel Asante Otchere, Tarek Omar Arbi Ganat, Jude Oghenerurie Ojero, Bennet Nii Tackie-Otoo, and Mohamed Yassir Taki. Application of gradient boosting regression model for the evaluation of feature selection techniques in improving reservoir characterisation predictions. *Journal of Petroleum Science and Engineering*, 208:109244, 2022.

[16] Amit Sharma, Neha Patel, and Rajesh Gupta. Enhancing sales efficiency with ai: Implementing random forest and logistic regression algorithms for lead scoring and qualification. *European Advanced AI Journal*, 11(8), 2022.

[17] Arpita Soni, Sahil Tripathi, Gautam Siddharth Kashyap, Manaswi Kulahara, Mohammad Anas Azeez, Zohaib Hasan Siddiqui, Nipun Joshi, and Jiechao Gao. Can we predict your next move without breaking your privacy? *arXiv preprint arXiv:2507.08843*, 2025.

[18] Fengli Zhang, Dana M Johnson, and Mark A Johnson. Development of a simulation model of biomass supply chain for biofuel production. *Renewable Energy*, 44:380–391, 2012.

Chapter 8

Predictive Maintenance in Critical Infrastructure Using Graph Neural Networks

Rafiq Ali

8.1 Introduction

Modern society is based on critical infrastructure systems (power grids, transportation networks, water distribution systems, and industrial facilities) [17]. This is necessary to guarantee their reliability and continuous operation to achieve economic stability [16], public safety [7], and environmental sustainability [21]. The conventional approaches to maintenance, such as reactive and preventive maintenance [14], tend to create high operation costs, unforeseen failures and poor use of resources. The growing access to sensor data and developments in artificial intelligence has led to the development of predictive maintenance as a useful technique to predict the occurrence of an equipment failure before it happens. Graph Neural Networks (GNNs) [23] have received an impressive amount of focus in this respect owing to the opportunity to model complicated system associations and spatial dependencies among system components. Figure 8.1 gives an overview of how GNNs can be used to predictively maintain critical infrastructure.

This chapter addresses the topic of using GNN-based platforms to monitor, diagnose, and predict failures in large-scale infrastructure systems. Section 8.3 of the paper covers data modeling methods, and learning architectures. Details of the deployment strategies are in Section 8.4 and the wheereas part in Section 8.5 covers the conclusion and some future directions of research.

DOI: 10.1201/9781003774815-8

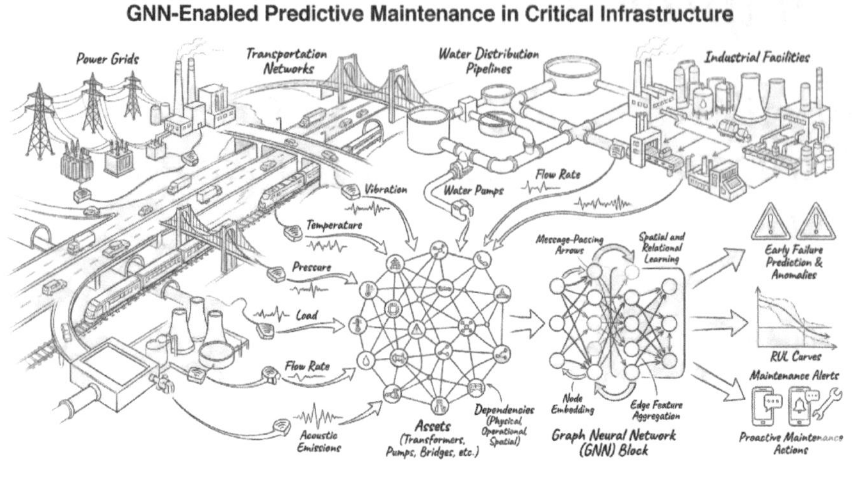

FIGURE 8.1: Overview of GNNs-based Predictive Maintenance in Critical Infrastructure Systems.

8.2 Graph-Based Modeling of Infrastructure Systems

There are some important infrastructure systems which are inherently interconnected, with components interacting through physical, logical and operational relationships [13]. Representing these systems as graphs does enables the explicit modeling of such dependencies where nodes correspond to assets like transformers, sensors, or machines and edges represent communication links, power flows, or functional relationships.

Graph construction which includes using heterogeneous data sources having sensor readings, maintenance logs, network topology and environmental information [20]. Temporal graphs further capture dynamic system behavior over time by enabling early detection of degradation patterns [9]. Feature engineering techniques have applied to encode operational states, failure indicators and contextual variables [6]. Figure 8.2 shows the process of transforming raw infrastructure data into structured graph representations suitable for GNN-based learning.

8.3 Graph Neural Network Architectures for Predictive Maintenance

GNNs extend traditional deep learning models by enabling information propagation across graph structures [23]. In predictive maintenance applications,

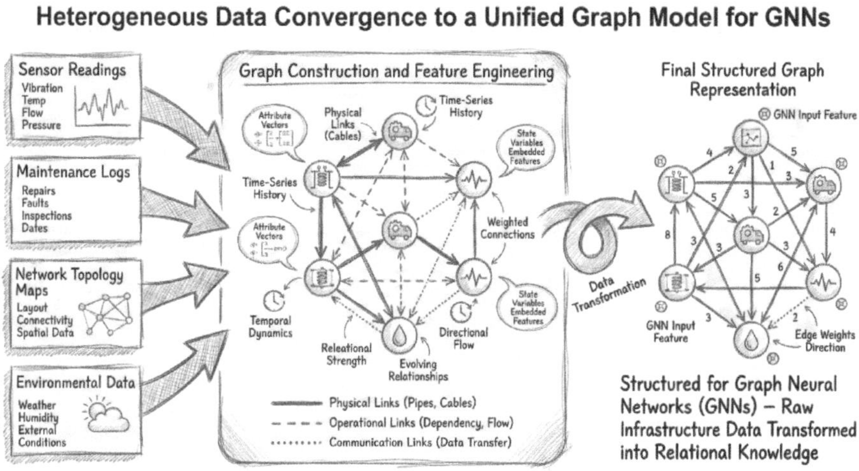

FIGURE 8.2: Graph-Based Modeling of Critical Infrastructure Systems from Heterogeneous Data Sources.

GNNs aggregate neighborhood information to learn spatial and temporal correlations among infrastructure components. Popular architectures include Graph Convolutional Networks (GCNs) [10], Graph Attention Networks (GATs) [22] and Temporal Graph Networks (TGNs) [18]. These models support tasks like anomaly detection [5] which remains useful life estimation [8], fault classification [25] and risk assessment [1]. Hybrid frameworks combining GNNs with recurrent or transformer-based models further increase temporal modeling capabilities.

Training GNNs requires carefully designed loss functions, regularization techniques and sampling strategies to handle large-scale graphs and class imbalance. Figure 8.3 presents a typical GNN-based predictive maintenance architecture by showing feature extraction, message passing and prediction layers.

8.4 Deployment Challenges and Applications

Deploying GNN-based predictive maintenance systems in real-world infrastructure environments presents different technical and organizational challenges [15]. These include data quality issues, system scalability [12] real-time inference requirements, cybersecurity risks and use of legacy platforms [2] (see Table 8.1).

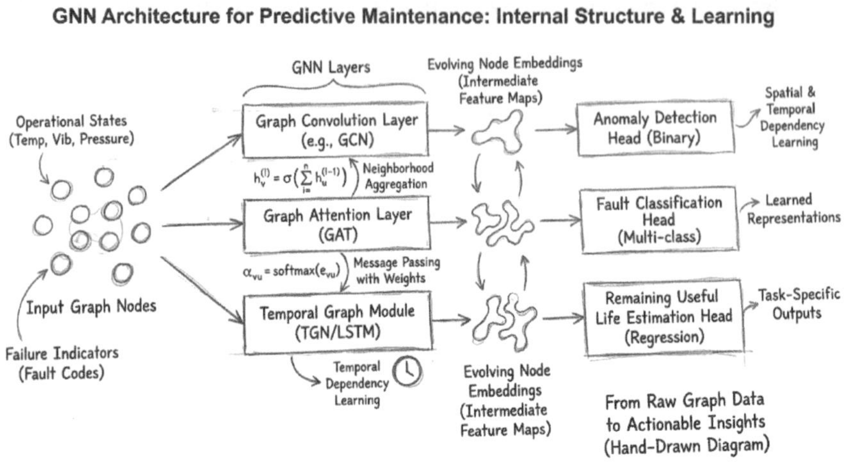

FIGURE 8.3: Architecture of GNNs for Predictive Maintenance Applications.

TABLE 8.1: Survey of Deployment Challenges and Applications in GNN-Based Predictive Maintenance

Category	Focus	Outcome
Data	Quality	Reliability
Scalability	Expansion	Efficiency
Latency	Inference	Responsiveness
Security	Protection	Resilience
Integration	Compatibility	Interoperability
Computation	Distribution	Optimization
Learning	Continuity	Adaptability
Compression	Reduction	Portability
Validation	Verification	Accuracy
Explainability	Transparency	Trust
Infrastructure	Modernization	Sustainability
Deployment	Automation	Stability
Applications	Industry	Impact

To solve these challenges there are some modern frameworks employ edge-cloud collaboration [3], distributed training [19], model compression [4] and continuous learning mechanisms. Strong validation and explainability techniques are also used to increase user trust and regulatory compliance [24]. Successful applications have been reported in power transmission networks, railway systems, manufacturing plants and smart cities [11]. Figure 8.4 shows

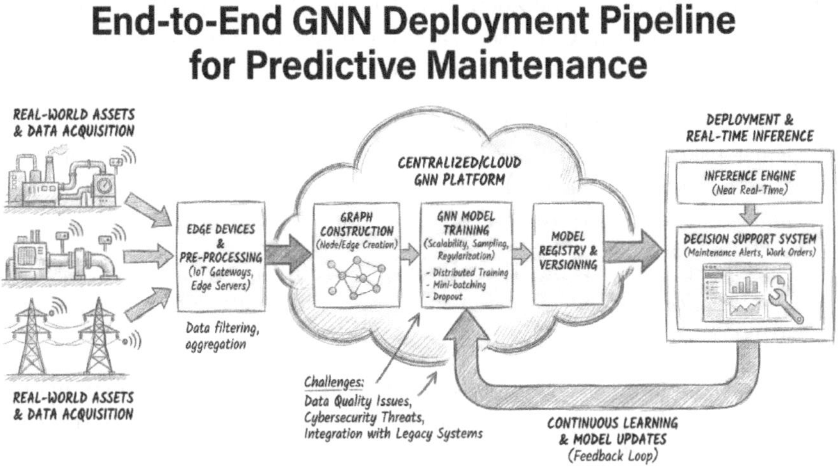

FIGURE 8.4: End-to-End Deployment Framework for GNN-Based Predictive Maintenance Systems.

the end-to-end deployment pipeline of GNN-based predictive maintenance systems from data acquisition to decision support.

8.5 Conclusion

In conclusion GNNs provide a powerful foundation for modeling complex infrastructure systems and enabling accurate predictive maintenance. With the help of capturing relational and temporal dependencies GNN-based frameworks improve failure prediction and asset management. Future research should focus on scalable learning algorithms, interpretable GNN models and standardized evaluation benchmarks to further advance the reliability and adoption of intelligent maintenance systems.

Bibliography

[1] Terje Aven. Risk assessment and risk management: Review of recent advances on their foundation. *European journal of operational research*, 253(1):1–13, 2016.

[2] Deval Bhamare, Maede Zolanvari, Aiman Erbad, Raj Jain, Khaled Khan, and Nader Meskin. Cybersecurity for industrial control systems: A survey. *computers & security*, 89:101677, 2020.

[3] Kun Cao, Shiyan Hu, Yang Shi, Armando Walter Colombo, Stamatis Karnouskos, and Xin Li. A survey on edge and edge-cloud computing assisted cyber-physical systems. *IEEE Transactions on Industrial Informatics*, 17(11):7806–7819, 2021.

[4] Yu Cheng, Duo Wang, Pan Zhou, and Tao Zhang. A survey of model compression and acceleration for deep neural networks. *arXiv preprint arXiv:1710.09282*, 2017.

[5] Ailin Deng and Bryan Hooi. Graph neural network-based anomaly detection in multivariate time series. In *Proceedings of the AAAI conference on artificial intelligence*, volume 35, pages 4027–4035, 2021.

[6] Vijay Govindarajan, Pratik Patel, Sahil Tripathi, Md Azizul Hoque, and Gautam Siddharth Kashyap. Magic-enhanced keyword prompting for zero-shot audio captioning with clip models. *arXiv preprint arXiv:2509.12591*, 2025.

[7] Y Haimes, J Lambert, and B Mahoney. Risk modeling, assessment, and management of interdependent critical infrastructures. *Charlottesburg: Center For Risk Management of Engineering Systems: University of Virginia*, 2000.

[8] Kai Huang, Guozhu Jia, Zeyu Jiao, Qun Wang, Feiyu Huang, and Yingjie Cai. Remaining useful life prediction through adaptive spatiotemporal graph neural network based on relationship mining for complex aviation equipment. *IEEE Sensors Journal*, 2025.

[9] Seyed Mehran Kazemi, Rishab Goel, Kshitij Jain, Ivan Kobyzev, Akshay Sethi, Peter Forsyth, and Pascal Poupart. Representation learning for dynamic graphs: A survey. *Journal of Machine Learning Research*, 21(70):1–73, 2020.

[10] TN Kipf. Semi-supervised classification with graph convolutional networks. *arXiv preprint arXiv:1609.02907*, 2016.

[11] Tong Liu and Fangyu Liu. Graph neural networks for evaluating the reliability and resilience of infrastructure systems: a systematic review of models, applications and future directions. *IEEE Access*, 2025.

[12] Hehuan Ma, Yu Rong, and Junzhou Huang. Graph neural networks: Scalability. In *Graph neural networks: foundations, frontiers, and applications*, pages 99–119. Springer, 2022.

[13] Jovica V Milanović and Wentao Zhu. Modeling of interconnected critical infrastructure systems using complex network theory. *IEEE Transactions on Smart Grid*, 9(5):4637–4648, 2017.

[14] R Keith Mobley. *An introduction to predictive maintenance.* Elsevier, 2002.

[15] Sai Munikoti, Deepesh Agarwal, Laya Das, Mahantesh Halappanavar, and Balasubramaniam Natarajan. Challenges and opportunities in deep reinforcement learning with graph neural networks: A comprehensive review of algorithms and applications. *IEEE transactions on neural networks and learning systems*, 35(11):15051–15071, 2023.

[16] Dmytro Myronchenko. Critical infrastructure and economic resilience. In *VIII International scientific and practical conference «Scientific Research as a Mechanism of Effective Human Development»(January 31-February 2, 2024) Sofia, Bulgaria, International Scientific Unity. 2024. 379 p.*, page 101, 2024.

[17] Steven M Rinaldi, James P Peerenboom, and Terrence K Kelly. Identifying, understanding, and analyzing critical infrastructure interdependencies. *IEEE control systems magazine*, 21(6):11–25, 2001.

[18] Emanuele Rossi, Ben Chamberlain, Fabrizio Frasca, Davide Eynard, Federico Monti, and Michael Bronstein. Temporal graph networks for deep learning on dynamic graphs. *arXiv preprint arXiv:2006.10637*, 2020.

[19] Yingxia Shao, Hongzheng Li, Xizhi Gu, Hongbo Yin, Yawen Li, Xupeng Miao, Wentao Zhang, Bin Cui, and Lei Chen. Distributed graph neural network training: A survey. *ACM Computing Surveys*, 56(8):1–39, 2024.

[20] Arpita Soni, Sahil Tripathi, Gautam Siddharth Kashyap, Manaswi Kulahara, Mohammad Anas Azeez, Zohaib Hasan Siddiqui, Nipun Joshi, and Jiechao Gao. Can we predict your next move without breaking your privacy? *arXiv preprint arXiv:2507.08843*, 2025.

[21] Bernhard Truffer, Harald Rohracher, Paula Kivimaa, Rob Raven, Floor Alkemade, Luis Carvalho, and Giuseppe Feola. A perspective on the future of sustainability transitions research. *Environmental Innovation and Societal Transitions*, 42:331–339, 2022.

[22] Petar Velickovic, Guillem Cucurull, Arantxa Casanova, Adriana Romero, Pietro Lio, Yoshua Bengio, et al. Graph attention networks. *stat*, 1050(20):10–48550, 2017.

[23] Zonghan Wu, Shirui Pan, Fengwen Chen, Guodong Long, Chengqi Zhang, and Philip S Yu. A comprehensive survey on graph neural networks. *IEEE transactions on neural networks and learning systems*, 32(1):4–24, 2020.

[24] John Zerilli. Explaining machine learning decisions. *Philosophy of Science*, 89(1):1–19, 2022.

[25] Zhewen Zhang and Lifeng Wu. Graph neural network-based bearing fault diagnosis using granger causality test. *Expert Systems with Applications*, 242:122827, 2024.

Chapter 9

AI for Smart Transportation: Road Safety, Traffic Flow, and Accident Prevention

Ebad Shabbir

9.1 Introduction

The modernization of urbanization and the increase in the number of vehicles has exerted unprecedented strain on the transportation infrastructures in the entire world [3]. There has been increasing traffic jams, accidents on the roads, and environmental pollution as a significant problem that impacts economic productivity, safety, and life quality in the cities and towns [11]. Conventional traffic management systems that strongly depend on fixed rules and hand operated monitoring are not always effective in identifying the dynamic traffic situation. Innovations in Artificial Intelligence (AI) [16], sensor technology, and communication networks have opened the way to creating smart transportation systems, which can monitor, predict and control in real-time [9]. Solutions that are AI-powered use information collected by cameras, GPS sensors, and connected cars, as well as road sensors to enhance situational awareness and decisions. Figure 9.1 gives a general picture of intelligent transportation systems with the use of AI and their main components.

This chapter discuss the aspect of using AI to improve road safety, traffic, and accident prevention. intel9 of the study talks about data-driven modeling, intelligent traffic management frameworks. Section 9.4 describes the challenges of the deployment, but Section 9.5 ends with some future research directions.

DOI: 10.1201/9781003774815-9

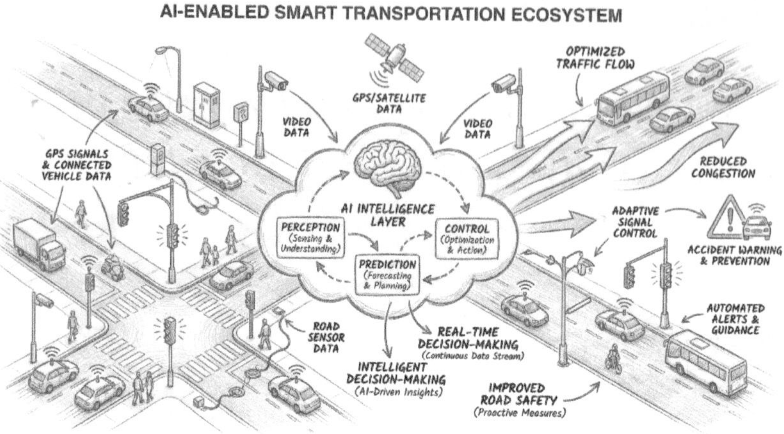

FIGURE 9.1: Overview of AI-Enabled Smart Transportation Systems and Their Core Components.

9.2 AI-Based Traffic Data Modeling and Sensing

Accurate traffic data collection and modeling form the foundation of intelligent transportation systems [1]. There are some modern smart transportation environments which uses heterogeneous sensing technologies [14] includes surveillance cameras [12], inductive loop detectors [2], LiDAR sensors [13] and vehicular communication units [10]. These devices generate large volumes of structured and unstructured data describing vehicle movements, traffic density and environmental conditions.

Machine learning techniques are used to process and analyze raw sensor data for traffic state estimation and anomaly detection [7]. Computer vision models enable vehicle detection, lane monitoring and pedestrian tracking while spatio-temporal learning methods capture patterns of traffic over time [18]. Data fusion techniques combine multiple data sources to increase reliability and robustness [5]. Figure 9.2 shows the process of traffic data acquisition, preprocessing and modeling in AI-driven transportation systems.

9.3 Intelligent Traffic Management and Accident Prevention

AI-powered traffic management systems aim to improve mobility and safety by optimizing signal control, route planning and incident response [6]. Reinforcement learning algorithms adapt traffic signal timings based on real-time

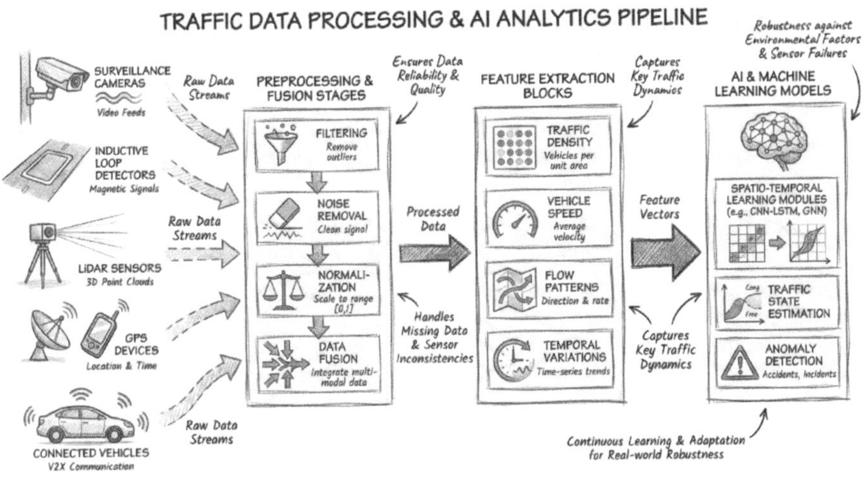

FIGURE 9.2: Traffic Data Acquisition, Preprocessing, and Modeling in AI-Driven Transportation Systems.

traffic conditions by reducing congestion and travel time [17]. Predictive models forecast traffic demand and identify potential type of bottlenecks.

For accident prevention DL models analyze video streams [4] and sensor data to detect risky driving behaviors like speeding, lane violations and distracted driving. Driver assistance systems and connected vehicle platforms provide real-time alerts to reduce collision risks [15]. Also anomaly detection frameworks identify abnormal traffic patterns associated with accidents or road hazards [19]. Figure 9.3 presents the architecture of AI-based traffic management and accident prevention systems by showing sensing, analytics and control layers.

9.4 Deployment Challenges and Applications

Although AI-based transportation systems have a set of potential benefits, there are several challenges to their deployment in the real world. These are data privacy, cost of infrastructure, scalability of the system, interoperability and cybersecurity risks and threats [8]. This further complicates deployment as it is necessary to ensure a reliable operation in changing weather conditions and during diverse traffic conditions.

In order to overcome these issues, the contemporary smart transportation systems embrace edge-cloud cooperation, federated learning, and distributed analytics [20]. Large-scale implementation is supported with the help of public- private partnerships and standardized communication protocols. It

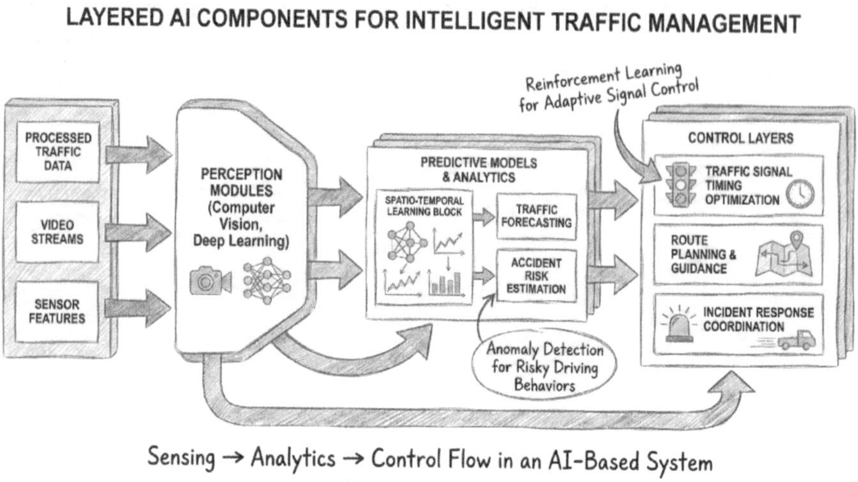

FIGURE 9.3: Architecture of AI-Based Traffic Management and Accident Prevention Systems.

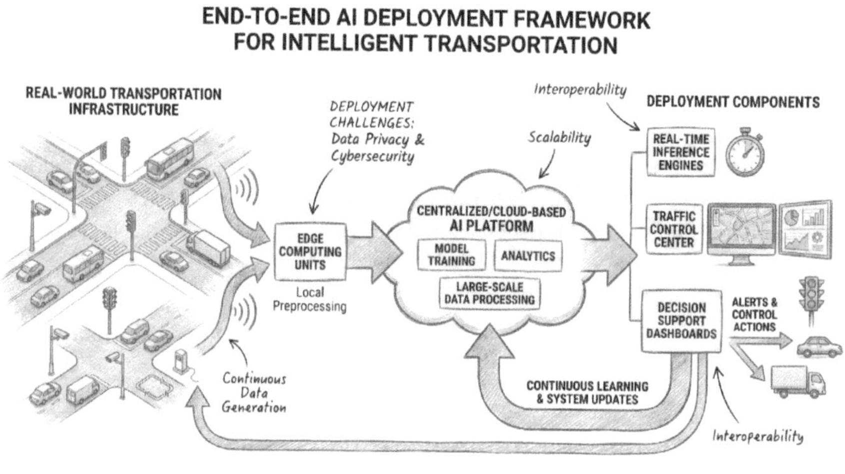

FIGURE 9.4: End-to-End Deployment Framework of AI-Driven Smart Transportation Systems.

has been reported to be successfully used in the intelligent traffic lights, automated toll applications, smart parking, and emergency response systems. The deployment model of the AI-based smart transportation systems is shown in Figure 9.4.

To address these challenges, modern smart transportation frameworks adopt edge-cloud collaboration, federated learning, and distributed analytics [20]. Public-private partnerships and standardized communication protocols facilitate large-scale implementation. Successful applications have been reported in intelligent traffic signal control, automated toll systems, smart parking, and emergency response management. Figure 9.4 illustrates the end-to-end deployment framework of AI-driven smart transportation systems.

9.5 Conclusion

To sum up, AI is a transformational tool in enhancing road safety, traffic efficiency and preventing accidents. AI-based transportation systems help to increase urban mobility and sustainability through the use of sophisticated sensing, data analytics, and adaptive control systems. The future research must be on explainable AI models, ethical governance and robust system architectures which will guarantee the success of intelligent transportation systems in the long run.

Bibliography

[1] Di Chen, Meixin Zhu, Hao Yang, Xuesong Wang, and Yinhai Wang. Data-driven traffic simulation: A comprehensive review. *IEEE Transactions on Intelligent Vehicles*, 9(4):4730–4748, 2024.

[2] Benjamin Coifman. Improved velocity estimation using single loop detectors. *Transportation Research Part A: Policy and Practice*, 35(10):863–880, 2001.

[3] Linlin Duan, Lulu Song, Wanjun Wang, Xiaomei Jian, Reinout Heijungs, and Wei-Qiang Chen. Urbanization inequality: evidence from vehicle ownership in chinese cities. *Humanities and social sciences communications*, 11(1):1–12, 2024.

[4] Vijay Govindarajan, Pratik Patel, Sahil Tripathi, Md Azizul Hoque, and Gautam Siddharth Kashyap. Magic-enhanced keyword prompting for zero-shot audio captioning with clip models. *arXiv preprint arXiv:2509.12591*, 2025.

[5] Bahador Khaleghi, Alaa Khamis, Fakhreddine O Karray, and Saiedeh N Razavi. Multisensor data fusion: A review of the state-of-the-art. *Information fusion*, 14(1):28–44, 2013.

[6] Li Li, Yisheng Lv, and Fei-Yue Wang. Traffic signal timing via deep reinforcement learning. *IEEE/CAA Journal of Automatica Sinica*, 3(3):247–254, 2016.

[7] Yisheng Lv, Yanjie Duan, Wenwen Kang, Zhengxi Li, and Fei-Yue Wang. Traffic flow prediction with big data: A deep learning approach. *Ieee transactions on intelligent transportation systems*, 16(2):865–873, 2014.

[8] Jonathan Petit and Steven E Shladover. Potential cyberattacks on automated vehicles. *IEEE Transactions on Intelligent transportation systems*, 16(2):546–556, 2014.

[9] Kashif Naseer Qureshi and Abdul Hanan Abdullah. A survey on intelligent transportation systems. *Middle-East Journal of Scientific Research*, 15(5):629–642, 2013.

[10] Bruno Ribeiro, Alexandre Santos, and Maria Joao Nicolau. A survey on vehicular communication technologies. *Intelligent Environments 2016*, pages 308–317, 2016.

[11] David Schrank, Bill Eisele, Tim Lomax, et al. Urban mobility report 2019. 2019.

[12] Arpita Soni, Sahil Tripathi, Gautam Siddharth Kashyap, Manaswi Kulahara, Mohammad Anas Azeez, Zohaib Hasan Siddiqui, Nipun Joshi, and Jiechao Gao. Can we predict your next move without breaking your privacy? *arXiv preprint arXiv:2507.08843*, 2025.

[13] Muhammad Sualeh and Gon-Woo Kim. Visual-lidar based 3d object detection and tracking for embedded systems. *IEEE Access*, 8:156285–156298, 2020.

[14] Anton Sysoev, Elena Khabibullina, Dmitry Kadasev, and Nikita Voronin. Heterogeneous data aggregation schemes to determine traffic flow parameters in regional intelligent transportation systems. *Transportation Research Procedia*, 45:507–513, 2020.

[15] Alireza Talebpour and Hani S Mahmassani. Influence of connected and autonomous vehicles on traffic flow stability and throughput. *Transportation research part C: emerging technologies*, 71:143–163, 2016.

[16] Eleni I Vlahogianni, Matthew G Karlaftis, and John C Golias. Short-term traffic forecasting: Where we are and where we're going. *Transportation Research Part C: Emerging Technologies*, 43:3–19, 2014.

[17] Hua Wei, Nan Xu, Huichu Zhang, Guanjie Zheng, Xinshi Zang, Chacha Chen, Weinan Zhang, Yanmin Zhu, Kai Xu, and Zhenhui Li. Colight: Learning network-level cooperation for traffic signal control. In *Proceedings of the 28th ACM international conference on information and knowledge management*, pages 1913–1922, 2019.

[18] Xueyan Yin, Genze Wu, Jinze Wei, Yanming Shen, Heng Qi, and Baocai Yin. Deep learning on traffic prediction: Methods, analysis, and future directions. *IEEE Transactions on Intelligent Transportation Systems*, 23(6):4927–4943, 2021.

[19] Bing Yu, Haoteng Yin, and Zhanxing Zhu. Spatio-temporal graph convolutional networks: A deep learning framework for traffic forecasting. *arXiv preprint arXiv:1709.04875*, 2017.

[20] Zhi Zhou, Xu Chen, En Li, Liekang Zeng, Ke Luo, and Junshan Zhang. Edge intelligence: Paving the last mile of artificial intelligence with edge computing. *Proceedings of the IEEE*, 107(8):1738–1762, 2019.

Part III

Decision Intelligence for Safer, Greener Infrastructure

Chapter 10

Securing Critical E-commerce Infrastructure: Unsupervised Learning and LLM-Enhanced Anomaly Detection for Web-Based Attacks

Nipun Joshi

10.1 Introduction

It has benefited global retail from electronic commerce that is more than $7 trillion a year and which is expected to surpass $8.1 trillion by 2026. These make platforms a necessity for economic infrastructure, processing billions of dollars of daily transactions. They're therefore ideal candidates for cyber threats that may disrupt operations, compromise data and undermine trust. Architectural complexity and attack sophistication are challenges to security [9, 35, 33, 17]. The modern e-commerce platforms have moved to distributed microservice architectures that have increased scale and agility but also added security. These are particular microservices that interface with APIs, increase the surface of attacks, and complicate inter-service vulnerabilities and authentication. The landscape of threats features low-level attacks designed to drain resources and nefarious, targeted attacks on data. Although DDoS attacks are fairly easy to detect [25], attacks such as Cross-Site Scripting (XSS) are particularly vulnerable. Both parameter tampering and the non-intrusive operation of a discrete device such as a keyboard [20] is uncomplicated, a difference that complicates volumetric analysis. The existing security protocols rely on signature-based [19], and rule-based detection. The network traffic analysis has been reported here [28, 32], as well as in a paper published in Science Advances. They are effective at coping with known threats, but they are particularly hard hit against large-scale, low-volume attacks targeting microservices. Base lines for traditional intrusion detection systems are difficult to establish. Performance measures lack context and rule-based systems do not

DOI: 10.1201/9781003774815-10

anticipate dynamic attack vectors. The longer-term implications of ephemeral deployments are further complicated.

Previous work has addressed unsupervised Machine Learning (ML) for anomaly detection [6]. This is consistent with the finding of anomalous patterns when combined with log data from the application data . But, even these methods can fail to detect even subtle attacks such as XSS and parameter tampering with a moderate level of precision. Therefore, there needs to be a better detection framework.

This chapter introduces an example of unsupervised ML coupled with LLM. Traditional approaches do not understand semantics, but LLMs excel in patterns of context. The proposed model strengthens security through statistical and semantic analysis. This integrated approach provides an extension of anomaly detection with LLM analysis of log data to detect subtle semantic irregularities.

The chapter is as follows in the following section, we will look at the theoretical backdrop. In Section 10.3, we shall look at related works. Section 10.4 describes the materials and methods. In Section 10.5, the experimental analysis is carried out. Section 10.6 discusses the case studies, and Section 10.7 wraps up the chapter with some conclusions and future work.

10.2 Theoretical Background

The modern e-commerce landscape has begun to take on a design shift that has led to the development of distributed microservices, which are scalable and agile. The microservices architecture is the development strategy where applications are developed as collections of tightly linked, independently deployable services that have specific business functions. e-commerce applications have increasingly come to rely on microservices architectures due to the complexity of systems which need to orchestrate functions like product catalogs, inventory management, payment processing, recommendation engines, user authentication and order fulfillment. E-commerce microservices environments tend to be the most critical architectural elements for resilient, scalable systems. API gateways are the primary gateway to external requests, redirecting traffic to appropriate services and meeting cross-cutting needs. Service discovery techniques can be used to automatically encrypt and locate service addresses, allowing components to communicate without the hard coded endpoints. Service meshes further extend this capability by providing better control, observation, and protection of traffic [36]. Containerization solutions such as Docker, package dependencies and microservices in light, portable boxes. Kubernetes and other container orchestration programs automate the deployment, scaling, and administration of containerized applications. Usually by CQRS, the database-per-service approach considers strategies to offer data via databases

TABLE 10.1: Survey of Unsupervised Machine Learning for Anomaly Detection

Method	Focus	Limitation
Autoencoder	Reconstruction	Thresholding
IsolationForest	Isolation	Sensitivity
OneClassSVM	Boundary	Scalability
Clustering	Grouping	Ambiguity
Density	Estimation	Sparsity
Statistical	Deviation	Rigidity
Ensemble	Aggregation	Complexity
Temporal	Sequencing	Drift
MetricBased	Monitoring	Semantics
FeatureBased	Engineering	Dependency

as a database per service approach. that allows each service access to a separate database store [21] and event sourcing to provide data consistency, but not tight coupling across services.

The microservices are distributed, leading to unique security problems. This broader attack surface is a central consequence because each microservice provides APIs to target. Communication gaps in services arise as services exchange information and could be exploited if the communication channels are not secured. To verify identities across service boundaries, authentication and authorization can be challenging, and often requires a tool such as OAuth 2.0. In the case of microservices, security implementations are inconsistent due to the polyglot nature of the environment because of organizational standards that are not adhered to. It is difficult to keep secrets as they need to be distributed and changed around. This observability gap also can hinder security monitoring because traditional tools can't reliably correlate events to reveal attacks across service boundaries [29].

10.3 Related Works

10.3.1 Machine Learning

Unsupervised anomaly detection is useful when protecting e-commerce environments where the available labeled attack information is limited and attack patterns change [6] (see Table 10.1). Unsupervised approaches provide baselines and deviations without pre-classified examples. This is especially relevant in the case of microservices where dynamic services and layered interactions have no static rules [27]. Unsupervised models can identify anomalies by identifying statistical outliers in performance measures and application logs [8]. It establishes the conceptual basis on three assumptions regarding the statistical nature of attacks: rarity, distinguishability, and deviation [11]. This suggests outliers in behavior that are just fractional to activity. This concept of distinguishability implies that behaviors have different features and thus there are separate feature spaces. The deviation principle holds that such deviations are deviations from norms. These principles inform algorithmic approaches with reconstruction, density, and isolation methods.

Auto encoders can encode and reconstruct normal data patterns, revealing anomalies through reconstruction errors. It comprises an encoder that reduces the dimensions of input, and a decoder that reconstructs the original input. It includes both an encoder that reduces the size of input and a decoder that reconstructs original input. The goal is to reduce reconstruction error so that the model can remember patterns. Anomalies are identified based on error thresholds. Isolation Forest is an ensemble design based on the principle that outliers are easier to isolate. Isolation Forest scales quickly, unlike density-based methods [2]. The algorithm constructs isolation trees and divides the feature space until the data points are isolated. This One-class Support Vector Machine approaches anomaly detection with a decision boundary around regular data points in feature space and the above-mentioned anomalies are detected beyond the boundary. The algorithm maps the input data to a higher-dimensional feature space using kernel functions, and then finds the hyperplane that maximizes the separation between the original and the transformed data points. A crucial finding from the previous studies is that unsupervised models vary in their effectiveness for all kinds of attacks [4]. Hence, volumes-based attacks such as GET floods [5], which vary significantly in performance measures, were consistently well-detected in all models. The brute force attacks of authentication [18], which cause small metric errors had good detection rates but with slightly less. But hegemonic attacks such as parameter tampering [19] and XSS are far more effective. That is, detection rates were much lower [20]. This difference explains a limitation of unsupervised metric based attacks; they fall short of effectiveness when attacks are not significantly impacting system performance. Parameter tampering and XSS are surgically precise and allow syntactic tweaking of parameters or client-side payload injections at low levels of performance signatures on the server. Statistical anomaly detection is useful for volumetric deviations but does not reveal the subtle manipulation patterns in normal performance parameters that are outside of any logic or security boundaries [7]. These problems show the need for wider approaches to the semantic gap in anomaly detection.

10.3.2 Large Language Models

LLMs represent a paradigm shift in computational linguistics [22], and has substantial potential for cybersecurity (see Table 10.2). Unlike NLP techniques that require statistics or set rules, LLMs employ transformer architectures on large text sources to gain further insights into language contexts, semantic

TABLE 10.2: Survey of Large Language Models for Security Log Analysis

Aspect	**Capability**	**Challenge**
Context	Understanding	Memory
Semantics	Interpretation	Ambiguity
Reasoning	Inference	Uncertainty
Detection	Identification	Latency
Integration	Hybridization	Overhead
Scalability	Processing	Resources
Privacy	Protection	Compliance
Adaptation	Learning	Drift
Validation	Verification	Reliability
Deployment	Automation	Governance

relationships, and domain specific patterns. This shift affords space to address the semantic gap in the traditional anomaly detection paradigm. These types of outliers are mainly statistically determined [6]. LLMs are promising in terms of enhancing detection, particularly in complex, semantic attacks which are hard to detect with conventional methods [34]. Similar log analyses in security have relied on pattern matching, regular expressions or simple statistical methods to detect text anomalies. Although computationally robust, these procedures lack the contextual knowledge needed to recognize complex attacks that are sufficiently correct but fail to meet semantic expectations. Initial NLP approaches utilized keyword extraction and clustering of similar messages, but they failed to detect manipulation attacks , such as parameter tampering or XSS, that fit nicely with legitimate traffic patterns [10].

It is also notable that application logs exhibit semantic disagreements that appear to be consistent with most statistical methods that use measures [1]. Likewise, XSS attacks inject script elements that are syntactically valid, yet are at the same time semantically deviant in some cases, necessitating contextual understanding in addition to pattern matching.

The implementation of LLMs in security is a concern that must be weighed against the computational and detection capabilities. The smaller task-specific models are more likely to be robust enough to analyze logs, and are less resource intensive. The size of context window is important as security sensitive patterns might overlap several log entries or appear to have similar events during the same time period. Even so, LLMs pose security challenges. Their computational strength also entails careful integration, particularly in the context of detection and resource use in high resolution e-commerce sites. Uncertainty is a factor in the probabilistic nature of LLM outputs which must be hedged by confidence thresholds and validation. Likewise, sensitive logs, which require careful data handling, pose security threats.

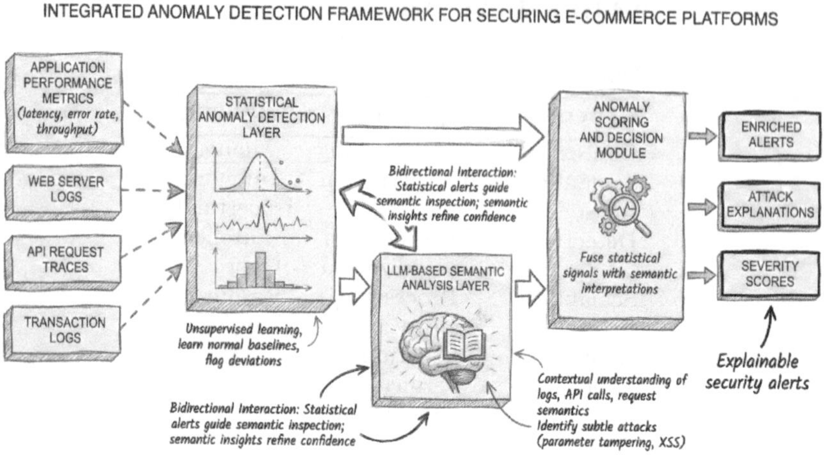

FIGURE 10.1: Architecture of the Proposed Integrated Anomaly Detection Framework Combining Statistical and LLM-Based Semantic Analysis.

10.4 Materials and Methods

The integrated framework contains four components (see Figure 10.1), namely data collection and preprocessing layer, unsupervised learning pipeline for statistical anomaly detection, LLM-based semantic analysis component, and an integration mechanism incorporating the learnings from both approaches. Each of the components addresses specific features of this detection problem, including the limitations in detecting simple attacks such as XSS and parameter tampering that maintain normal performance. The data collection layer collects performance measures and application logs from distributed microservices. Application logs provide additional information on semantic behavior as it records interactions and events that may not be reported in performance metrics. We collect performance measures in time-series format for querying and analysis, and the EFK stack includes and identifies log streams for further analysis using Elasticsearch, Fluentd, Kibana. The preprocessing step prepares the data for analysis by cleaning out missing values, adjusting dimensions and determining differential values to highlight changes rather than absolute values. Log preprocessing standardized formats across services and extracts structured information from unstructured records. A major feature is contextual aggregation of logs that brings together similar entries across service boundaries to provide a full picture of the interaction between specific microservices.

The unsupervised detection pipeline builds on this by merging multiple model types to maximize their advantages. In the pipeline, three major model types are used; reconstruction-based models, such as autoencoders; isolation-based models, such as isolation forest; and boundary-based models, such as One-Class SVM. Each model type is able to identify different anomaly patterns with autoencoders being effective for volumetric attacks and Isolation Forest being particularly sensitive for identifying attack instances across categories. The autoencoder part employs a special architecture which is designed to handle the high-dimensional time series data of microservices environments. It does not require an autoencoder, but rather takes temporal data as a sliding window which captures sequential patterns in metrics. This refinement addresses a limitation in previous research where time relationships between measures were not used fully. In essence, there are several hidden layers of autoencoder architecture whose dimension decreases and increases, creating a bottleneck that allows the model to learn effective representations of normal behavior patterns. Training is only performed on average operation data, and there are reconstruction error thresholds that specify deviations indicating possible attacks. Isolation Forest also provides complementary detection by correctly identifying outliers in large dimensions. Unlike density-based approaches, which struggle with the curse of dimensionality, Isolation Forest effectively adapts to the richness of data generated by microservices environments.

The new approach addresses the fundamental limitation of existing methods of detecting semantically complex attacks through the LLM-enhanced analysis component. This component uses the contextual knowledge of LLMs to identify suspicious patterns in application logs that might otherwise appear normal using statistical techniques. In this approach, the method takes a multi-step approach to address computational limitations while being sufficiently sensitive to detection. The first is a thin language model based on domain-specific patterns associated with security in the e-commerce application logs. Logs in this model are checked for semantic anomalies such as missing parameter values, unusual sequence of requests, or abnormal patterns of data that are not consistent with behavioral patterns. This fine-tuning is performed using both real application logs and artificial attacks that enable the model to distinguish between legitimate and potentially malicious behavior without providing any clear definitions of rules. Using a more sophisticated language model with larger context windows for suspicious logs, a second step is performed in depth. It deals with the contents of the individual log entries and the relationship between the log entries across service boundaries in this model. It can also identify semantic contradictions arising from more complex attacks, such as parameter tampering or XSS injection, by analyzing the expected flow of transactions between microservices for legitimate transactions. This is especially useful for identifying attacks that have similar performance characteristics but are outside of the application logic or security boundary.

This layer is a critical link between the detection of statistical anomalies and the semantic analysis, where the results from both methods are synthesized together to make detection capabilities more general. Instead of treating them as a set of parallel systems with separate outputs, the framework has developed a particularly complex correlation framework that tracks correlation between metric and semantic anomalies across time and service boundaries. This is further incorporated into an integrated decision model with varying levels of detection confidence, anomaly severity and contextual variables to provide relevant security information. At the first level, the framework blends metrics and log-based semantic anomalies that occur within configurable time frames and service boundaries. This temporal and spatial correlation allows for the differentiation between anomalies that are non-uniform and that may be false positive and patterns that may reflect actual attack activity. The correlation engine gives confidence scores for the strength of the association between anomaly types, and higher confidence scores for patterns consistent with the known attack signatures for both performance impact and semantic properties. The second level provides an explanational decision framework which transposes detection data into useful insights for security analysts. The system does not just identify anomalies, but provides contextual information on the nature of anomalies detected, the impact they may have, and what evidence supports detection decisions. This explication is especially important in the case of security operations in production settings where analysts need to quickly assess warnings and assess how to respond. Each anomaly spawned in the framework is explained as follows: which services were affected, what metrics or log patterns were observed at the time of detection, and possibly type attacks based on similarity to patterns.

10.5 Experimental Analysis

In order to evaluate the integrated anomaly detection methodology, a number of attacks were tested for the presence of the integrated anomaly detection model including unsupervised attacks as well as experiments on the improved LLM-integrated approach. Tables 10.3–10.5 compare baseline unsupervised models and the proposed LLM-integrated framework on the measures of overall and attack-wise detection. This experiment utilized a microservices application with 41 connected services on a Kubernetes cluster. Performance measures and application logs were collected over four weeks to establish baselines of normal behavior before introducing simulated attack situations. These attack scenarios included volumetric attacks (GET floods), authentication attacks (brute forcing), parameter manipulation (tampering with request parameters) and client-side injection (Cross-Site Scripting, i.e. XSS). These types are a mixture of threats, ranging from high levels of anomalies to sub-

TABLE 10.3: Performance of Baseline Unsupervised Anomaly Detection Models

Model	Accuracy (%)	Precision (%)	Recall (%)
Autoencoder	75	74	76
Isolation Forest	72	55	96
One-Class SVM	70	49	94

TABLE 10.4: Performance of LLM-Integrated Anomaly Detection Framework

Metric	Accuracy (%)	Precision (%)	Recall (%)
Overall Performance	83	78	85
Attack Detection	81	80	84
Normal Traffic	86	82	87

TABLE 10.5: Attack-Wise Detection Performance of the Integrated Framework

Attack Type	Detected Instances	Detection Rate (%)
GET Flood	16/16	100
Brute Force	14/16	87.5
Parameter Tampering	2/2	100
XSS	13/15	86.7

tle semantic manipulations. First-line experiments using unsupervised models without enhancement of LLM confirm the findings of the previous study and reveal new limitations. Autoencoder models were moderately accurate (75%) with good recall and precision for normal and attack classes. But, the overall comparison was clear on attack type. GET flood attacks were consistently identified with high accuracy (16/16 instances) and authentication brute force attacks were also quite effective (10-15/16 instances). While the parameter tampering and XSS attacks were hard to detect, some models were very poor, such as 0/15 XSS detections using too complete autoencoders. Isolation Forest showed high sensitivity to attacks (96% recall), but was too sensitive (55% recall) and produced numerous false positives. One Class SVM was similarly characterized, recalling 94% but lacking 49% accuracy. These results showed that metric-based anomaly detection is limited to sophisticated attacks that have stable performance.

LLM-based improved log analysis gave a marked improvement in performance for all dimensions. The integrated framework was 8% more accurate than the best-performing baseline model. The detection of very subtle attacks received significant enhancements. Previously detection rates for XSS attacks were as low as 0/15 instances, but the integrated framework identified 13/15 instances with reasonably good accuracy. This improvement was due to the LLM's ability to detect semantic anomalies in logs that appeared to be normal to statistical reasoning. Attacks on parameter tampering enabled detection over 0–2/2 instances with higher confidence scores. False positive reduction improved significantly. For example, 78% of attacks were classified accurately using the combined approach compared to 55%-70% of attacks using the traditional model. This directly addressed operational problems in production areas where excessive false positives can lead to alert fatigue and overlooked real threats. Analysis by attack category revealed specific patterns of how the combined strategy improved detection. Plus, for volume attacks such as GET floods, conventional, unsupervised approaches had already demonstrated near-perfect detection (15–16/16). This high performance was retained through the integration and additional contextual information emerged via LLM analysis of log patterns. LLM was used in the detection of authentication brute force attacks that had much more precise detection because it reflected semantic patterns in authentication failure logs. Identifiers made by legitimate authentication were not infected by coordinated brute force attacks; hence false positives were reduced while providing high sensitivity.

10.6 Discussions

The integrated anomaly detection paradigm stretches beyond theoretical considerations to practical applications in the real world of e-commerce. A case study of this approach describes how it addresses security threats in a payment processing system that handles sensitive financial transactions.

10.6.1 Case chapter 1

This framework was used by a financial technology provider that processes more than 5 million daily transactions in order to strengthen security controls in its microservices environment. This payment infrastructure consists of approximately 35 microservices and processes payment authorization, fraud screening, transaction routing, settlement processing, and customer notification. Each service ran its own in containerized environments with Kubernetes and acted as an open architecture with big attack surfaces. The framework was designed to detect sophisticated attacks centered on transaction processing or authorization workflows, patterns that conventional detection systems were far too stymied. This deployment relied on existing monitoring tools: Prometheus was used to track performance and EFK stack to store application logs. Contextual aggregation of logs and measures for specific flow of transactions giving in-depth visibility across service boundaries. The unsupervised detection pipeline utilized autoencoders to identify synchronized transactional phases, and the identification of abnormal processing sequences as

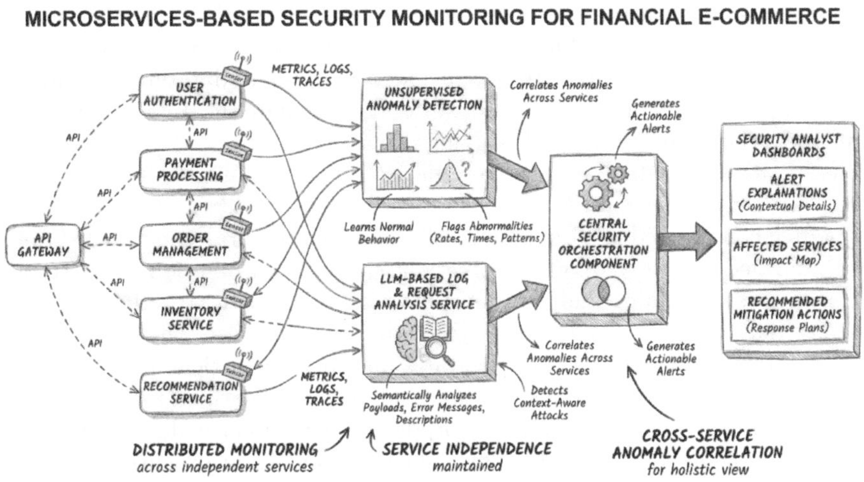

FIGURE 10.2: Microservices-Based Security Monitoring and Anomaly Detection Architecture for the Financial Transaction Case Study.

evidence of manipulation was essential. The enhanced log analysis aspect of LLM was an important advantage for security since it found patterns of semantics that reflected parameter tampering, redirect manipulation or power bypass. Transaction-centric correlation reduced false positives and provided a common solution to the anomaly across services involved in particular payment flows. Operational results demonstrated significant improvements in the detection of sophisticated attacks. During the assessment period the framework found 18 examples of parameter tampering for systems previously not included in the model. Detection latencies averaged 175 ms, enabling blockages of transactions earlier to protect the taxpayer from losses. True positive rates also decreased by 67%, reducing the operational burden and still having high detection sensitivity. In addition, the LLM component found a previously untold attack pattern of invisibility of authorization parameters across multiple services [30, 31, 3, 26, 14, 12, 13, 16, 15, 24, 23]. Figure 10.2 illustrates the security monitoring architecture and anomaly detection workflow using microservices in the financial transaction case study.

10.7 Conclusions

This cohesive framework provides considerable improvements in detection capabilities for multiple attack vectors, but there are also several challenges and limitations that warrant further testing. This framework is effective against existing attack patterns but may have difficulty with zero-day exploits that use new mechanisms that are not included in existing datasets. The limitation applies to all anomaly-based systems, although the LLM components mitigate this limitation by identifying suspicious patterns that might reflect unseen attack strategies. Plus, complicated language models require an increasing computational load and can present deployment challenges in highly resource-constrained environments, offering architectural choices that must be managed carefully between detection features and functional constraints. Unsupervised learning and LLM outputs are probabilistic, resulting in uncertainty that must be controlled for through threshold selection and validation to maintain sufficient false positive rates and identify critical attacks.

There are several promising directions for further investigation; multimodal anomaly detection that incorporates performance data, application logs, user behaviour, network flow behavior, and code execution traces to design complete security profiles. This could increase the detection of complex attacks at various scales, but it will have no footprint in a single area. Another route that seems promising is reinforcement learning strategies that ameliorate dynamic threshold adjustment, i.e., adaptive security systems that modifiably adjust detection sensitiveities on changing threats and traffic patterns without manual reconfiguration. Federated learning practices allow for the sharing of model training across organizations, without the complexities of storing and analyzing sensitive data. These techniques can be used to create robust detection models on many different attack observation sets and with the data being confidential.

Bibliography

[1] Sawsan AbdulRahman, Hanine Tout, Hakima Ould-Slimane, Azzam Mourad, Chamseddine Talhi, and Mohsen Guizani. A survey on federated learning: The journey from centralized to distributed on-site learning and beyond. *IEEE Internet of Things Journal*, 8(7):5476–5497, 2020.

[2] Hadeer Ahmed, Issa Traore, and Sherif Saad. Detection of online fake news using n-gram analysis and machine learning techniques. In *International conference on intelligent, secure, and dependable systems in distributed and cloud environments*, pages 127–138. Springer, 2017.

[3] Venu Madhav Aragani. Evaluating reinforcement learning agents for portfolio management. In *2025 Fifth International Conference on Advances in Electrical, Computing, Communication and Sustainable Technologies (ICAECT)*, pages 1–6. IEEE, 2025.

[4] Divyanshu Awasthi and Vinay Kumar Srivastava. Hessenberg decomposition-based medical image watermarking with its performance comparison by particle swarm and jaya optimization algorithms for different wavelets and its authentication using aes. *Circuits, Systems, and Signal Processing*, 42(8):4953–4984, 2023.

[5] Sunny Behal and Krishan Kumar. Characterization and comparison of ddos attack tools and traffic generators: A review. *Int. J. Netw. Secur.*, 19(3):383–393, 2017.

[6] Varun Chandola, Arindam Banerjee, and Vipin Kumar. Anomaly detection: A survey. *ACM computing surveys (CSUR)*, 41(3):1–58, 2009.

[7] Kevin Delmolino, Mitchell Arnett, Ahmed Kosba, Andrew Miller, and Elaine Shi. Step by step towards creating a safe smart contract: Lessons and insights from a cryptocurrency lab. In *International conference on financial cryptography and data security*, pages 79–94. Springer, 2016.

[8] Min Du, Feifei Li, Guineng Zheng, and Vivek Srikumar. Deeplog: Anomaly detection and diagnosis from system logs through deep learning. In *Proceedings of the 2017 ACM SIGSAC conference on computer and communications security*, pages 1285–1298, 2017.

[9] Rudresh Dwivedi and Somnath Dey. A novel hybrid score level and decision level fusion scheme for cancelable multi-biometric verification. *arXiv preprint arXiv:1805.10433*, 2018.

[10] Ashish Ghosh, Debasrita Chakraborty, and Anwesha Law. Artificial intelligence in internet of things. *CAAI Transactions on Intelligence Technology*, 3(4):208–218, 2018.

[11] Oded Goldreich and Rafail Ostrovsky. Software protection and simulation on oblivious rams. *Journal of the ACM (JACM)*, 43(3):431–473, 1996.

[12] Srinivas R Gottimukkala. Applying the multifractal model of asset returns (MMAR) to financial markets: Insights and limitations. In *2025 10th International Conference on Signal Processing and Communication (ICSC)*, pages 277–282. IEEE, 2025.

[13] Srinivas R Gottimukkala. Automating portfolio replication with stock market algorithms. In *2025 8th International Conference on Circuit, Power & Computing Technologies (ICCPCT)*, pages 973–978. IEEE, 2025.

[14] Srinivas R Gottimukkala. A computational approach to replicating correlated portfolios using algorithmic insights from stock market dynamics. In *International Conference on Information and Communication Technology for Intelligent Systems*, pages 265–275. Springer, 2025.

[15] Srinivas R Gottimukkala. Optimizing exotic option pricing: Monte carlo simulation and variance reduction techniques. In *International Conference on Information and Communication Technology for Intelligent Systems*, pages 277–288. Springer, 2025.

[16] Srinivas Raju Gottimukkala. Evaluating the impact of fed and domestic monetary policies on long-term government bond yields. 2024.

[17] Vijay Govindarajan, Pratik Patel, Sahil Tripathi, Md Azizul Hoque, and Gautam Siddharth Kashyap. Magic-enhanced keyword prompting for zero-shot audio captioning with clip models. *arXiv preprint arXiv:2509.12591*, 2025.

[18] Martin Husák, Jana Komárková, Elias Bou-Harb, and Pavel Čeleda. Survey of attack projection, prediction, and forecasting in cyber security. *IEEE Communications Surveys & Tutorials*, 21(1):640–660, 2018.

[19] Muhammad Idris, Iwan Syarif, and Idris Winarno. Development of vulnerable web application based on owasp api security risks. In *2021 International Electronics Symposium (IES)*, pages 190–194. IEEE, 2021.

[20] Martin Johns, Björn Engelmann, and Joachim Posegga. Xssds: Server-side detection of cross-site scripting attacks. In *2008 Annual Computer Security Applications Conference (ACSAC)*, pages 335–344. IEEE, 2008.

[21] Oleksandr Lytvynov and Dmytro Hruzin. Decision-making on command query responsibility segregation with event sourcing architectural variations. *Technology audit and production reserves*, 4(84), 2025.

[22] Chenyang Lyu, Zefeng Du, Jitao Xu, Yitao Duan, Minghao Wu, Teresa Lynn, Alham Fikri Aji, Derek F Wong, and Longyue Wang. A paradigm shift: The future of machine translation lies with large language models. In *Proceedings of the 2024 Joint International Conference on Computational Linguistics, Language Resources and Evaluation (LREC-COLING 2024)*, pages 1339–1352, 2024.

[23] Satish Mandavalli. Enhancing dengue outbreak predictions using machine learning: A comparative analysis of models. 2024.

[24] Satish Mandavalli. Enhancing crop image classification: Comparative analysis of augmentation techniques for small datasets. In *International Conference on Information and Communication Technology for Intelligent Systems*, pages 591–604. Springer, 2025.

[25] Jelena Mirkovic and Peter Reiher. A taxonomy of ddos attack and ddos defense mechanisms. *ACM SIGCOMM Computer Communication Review*, 34(2):39–53, 2004.

[26] Lakshmi Narasimha Raju Mudunuri, Praveen Kumar Maroju, and Venu Madhav Aragani. Leveraging nlp-driven sentiment analysis for enhancing decision-making in supply chain management. In *2025 Fifth International Conference on Advances in Electrical, Computing, Communication and Sustainable Technologies (ICAECT)*, pages 1–6. IEEE, 2025.

[27] Sneha Padhiar and Ritesh Patel. Outside the closed world: On using machine learning for network intrusion detection. In *International Conference on Information and Communication Technology for Intelligent Systems*, pages 265–270. Springer, 2023.

[28] Martin Roesch et al. Snort: Lightweight intrusion detection for networks. In *Lisa*, volume 99, pages 229–238, 1999.

[29] D Santhadevi and B Janet. Light weight gradient ensemble model for detecting network attack at the edge of the iot network. *International Journal of Information Technology*, 15(1):29–38, 2023.

[30] Parth Saxena. Evaluating machine learning algorithms for effective credit card fraud detection. In *2025 10th International Conference on Signal Processing and Communication (ICSC)*, pages 375–380. IEEE, 2025.

[31] Parth Saxena and Janak Devashish. Sentiment analysis and topic modeling of patient reviews on psychiatric medications using cloud-based architecture. In *2025 International Conference on Computational, Communication and Information Technology (ICCCIT)*, pages 892–897. IEEE, 2025.

[32] Peter Warren Singer and Allan Friedman. *Cybersecurity and cyberwar: What everyone needs to know®*. Oxford University Press, 2013.

[33] Arpita Soni, Sahil Tripathi, Gautam Siddharth Kashyap, Manaswi Kulahara, Mohammad Anas Azeez, Zohaib Hasan Siddiqui, Nipun Joshi, and Jiechao Gao. Can we predict your next move without breaking your privacy? *arXiv preprint arXiv:2507.08843*, 2025.

[34] GSDMA Sreenu and Saleem Durai. Intelligent video surveillance: a review through deep learning techniques for crowd analysis. *Journal of Big Data*, 6(1):1–27, 2019.

[35] Abhinav Srivastava, Amlan Kundu, Shamik Sural, and Arun Majumdar. Credit card fraud detection using hidden markov model. *IEEE Transactions on dependable and secure computing*, 5(1):37–48, 2008.

[36] David Tilman, Christian Balzer, Jason Hill, and Belinda L Befort. Global food demand and the sustainable intensification of agriculture. *Proceedings of the national academy of sciences*, 108(50):20260–20264, 2011.

Chapter 11

Advancing Software Quality with Large Language Models, Deep Learning, and Cloud Infrastructure

Krishna Gandhi, Pankaj Verma, Vijay Govindarajan, and Raj Sonani

11.1 Introduction

Software Quality Automation (SQA) is a term that refers to the degree to which a program or piece of software meets the needs both of the user and of the developer [8]. It addresses usability, security, reliability, performance and functionality. Quality software works smooth, produces exact results and gives a clean user experience. Poor software, with security vulnerabilities, numerous crashes and operational inefficiency, can harm businesses and end-users [11]. Software must be reliable and can be properly maintained, and can function correctly over time without errors [26]. Software that is reliable reduces the downtime, prevents data corruption, and allows a more comfortable user experience. Another is the performance of the software system or the speed with which it performs tasks in different situations. Poorly optimized software wastes resources, slows response times and scales poorly, ultimately reducing productivity and user satisfaction. In today's cyberworld, quality software is important [5]. Software vulnerabilities can be source of data breaches, financial damages and lawsuits [39, 10]. Software quality needs cybersecurity standards, vulnerability tests and secure coding practices. Another consideration is maintainability. Written code, organized and documented, enables future changes, bug fixes and improvements to be implemented without rework.

High quality software is required in healthcare [28], banking [29], and aviation [46]. Here error can be extremely serious. This kind of software failure can be fatal, costly, or unimaginable. A miserly medical imaging system may produce inaccurate scan results [13], and incorrect diagnoses and treatment options. Quality software in healthcare is dependent upon rigorous testing,

DOI: 10.1201/9781003774815-11

compliance with HIPAA and FDA regulations, and vulnerability testing [27]. Financial modeling [28], stock trading [7], fraud detection [30], and banking [1] use software extensively. Even a small software glitch can cause fund inconsistencies, illegal transactions, or security breaches in banking apps. After the 2010 "Flash Crash" that destroyed almost $1 trillion in market value in minutes due to an algorithmic trading error [15], reliable financial software is essential. Financial software quality requires extensive performance testing to manage high transaction volumes, PCI-DSS and GDPR compliance, and strict security measures.

The chapter is as follows; in the next section, we will look at the theoretical backdrop. Section 11.3 discusses the use of ML for quality assurance. Section 11.4 discusses the use of LLMs to improve software quality. In Section 11.5, ML in software testing is discussed. Section 11.6 discusses DL-based flaw identification, and Section 11.7 wraps up the research with some conclusions and future work.

11.2 Theoretical Background

SQA has grown due to the need for more reliable, efficient, and scalable testing methods [36]. There was need for ML, rule-based testing [2], and automation [14] because old software systems didn't exist without manual testing. This change has led to increased reliability, testing efficiency, and defect detection of software [35]. Human testers conducted tests by working with software programs, finding bugs, and writing documentation for developers. This process required detailed test scripts, multiple testing environments and manual output verification to ensure accuracy. Manual testing was effective for small applications but not for large systems due to a few constraints. The manual labor was tedious. Testing lasted longer and cost more because each test case was run separately. As the software applications grew and became more functional, the number of test cases needed to pass the full test became exponential, rendering manual testing unworkable. Bugs may occur that manual testers forget to see due to fatigue, inconsistency or cognitive bias. Bugs of important quality would less likely hit the production stage, leading to software failures or costly fixes. Rule-based defect detection systems were more efficient than manual testing [45]. These systems were identified through heuristics and predefined rules with outflows of memory, buffer overflows, syntax errors, and performance bottlenecks [41]. Checkstyle, SonarQube, PMD, and other static analysis tools check source code for coding standards and best practices violations. Automation provided by these tools with feedback helped developers to address bugs before they started using the software [42]. Dynamic analysis was developed to identify runtime errors that static analysis did not detect. Rule-based testing systems involved less human intervention in software quality assurance. The rules based system, though effective, was bleak. Their rules only endorsed known software vulnerabilities rather than new ones. The many false positives in rule-based systems forced engineers to look for non-defects. These limitations necessitated better-informed, adaptive solutions to quality assurance that learned from mistakes and predicted future concerns.

11.3 Automation via Machine Learning in Quality Assurance

Test automation revolutionized software quality assurance to overcome the inefficiencies of manual and rule-based testing. Selenium, JUnit, and TestNG enabled testers to write scripts that ran in test environments without human involvement [31]. This significantly increased the test speed, consistency and scale of large-scale software projects. Automation of tests would allow for thousands of tests to be performed parallel, allowing for quicker identification of software problems. Unlike manual testing, which required time and resources, automation was more accurate in determining the functionality of the software. Automation enabled continuous testing throughout the software development life cycle avoiding costly rework. Traditional test automation still required human script creation and maintenance [25]. Application feature updates required the maintenance of test scripts. Better was programmatic automated testing that didn't find a solution for software problems. This led to the use of ML in software testing, greatly increasing the quality assurance [23]. ML models can also identify patterns in the defect data and predict software issues without rule-writing, versus rule based systems. ML prediction began with decision trees [9], SVMs [4], and random forests [38].These models helped software developers identify codebase challenges by assessing code complexity, defect frequency, and commit history. The ML-based anomaly detection methods identified software behaviour problems that had been ignored by traditional testing [24]. The major contribution of ML to software quality assurance is self-healing automated frameworks. In reaction to user interface and functionality changes in these frameworks, ML updated test scripts without human intervention. This made testing procedures less interrupted and automated more software update-resistant. DL techniques like RNNs [19], CNNs [3], and neural networks [33] helped AI-driven software testing tools analyse complicated software structures, discover tiny bugs, and offer automatic problem remedies. The shift from rule-based to ML-driven quality assurance has enabled AI-driven software testing. Figure 11.1 illustrates the architecture of machine learning-driven automated testing and quality assurance systems.

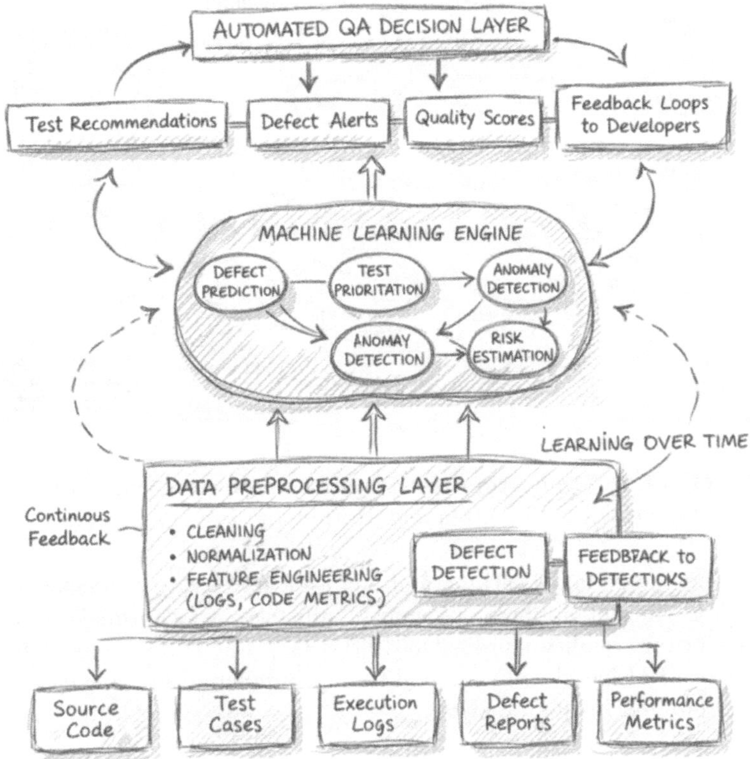

FIGURE 11.1: Architecture of Machine Learning-Driven Automated Testing and Quality Assurance Systems.

11.4 Large Language Models for Software Quality Improvement

LLMs and DL models transformed software quality assurance. Models that understand and generate human-like text are useful for automated code reviews [22], documentation [43], bug classification [32], and more. Models like Codex and GPT-4o can evaluate code, optimize it, and fill documentation gaps. DL models are trained on large software repositories to improve maintainability, refactor inefficient code, and identify coding patterns. Integration of AI, ML and LLMs into software quality assurance increases reliability, testing time

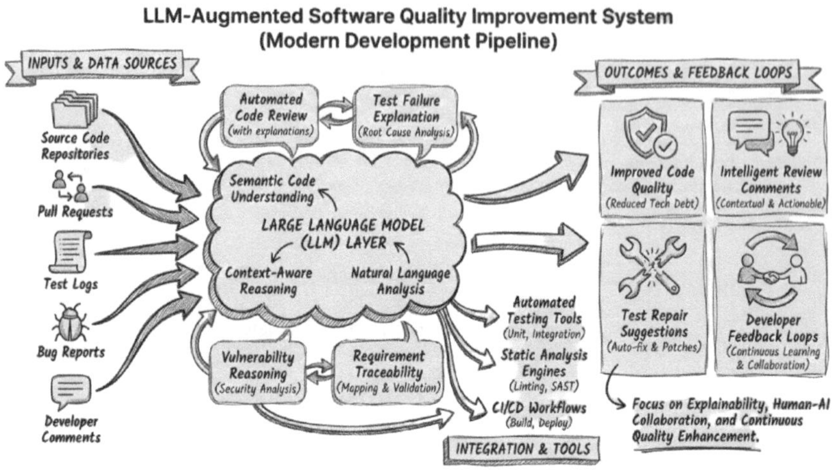

FIGURE 11.2: Architecture of LLM-Driven Software Quality Improvement Systems Integrated with Automated Testing and Review Pipelines.

and defects detection. Learning and adapting AI based approaches will lead to robust, efficient applications across industries, which will be critical for future software quality [16] [32][40][12][17] [37]. Figure 11.2 shows the architecture of LLM-based software quality improvement systems and integration with automated testing pipelines.

11.5 Machine Learning in Software Testing

ML has transformed software quality assurance, test case optimization, and defect detection as shown in Figure 11.3. ML models were studied to better software testing due to large datasets and processing capacity. ML algorithms could learn from defect data, adapt to new patterns, and predict software quality, unlike rule-based methods [6] [34] [21] .Early ML models like Decision Trees (DTs), Random Forests (RFs), and SVMs used data to detect issues, classify errors, and optimize testing. These methods dramatically enhanced software teams' bug prediction and avoidance before production, making software more secure and reliable. In software testing, ML predicted defects. Training algorithms identified bug-prone code areas. DTs and RFs were often employed to identify harmful software. These models could analyse code complexity, commit history, developer activity, etc. Software modules were graded by defect likelihood in the models' decision hierarchies [20]. DTs and RFs learnt from issue reports and software updates to assist teams test the most vulnerable regions, enhancing software quality. Bug classification

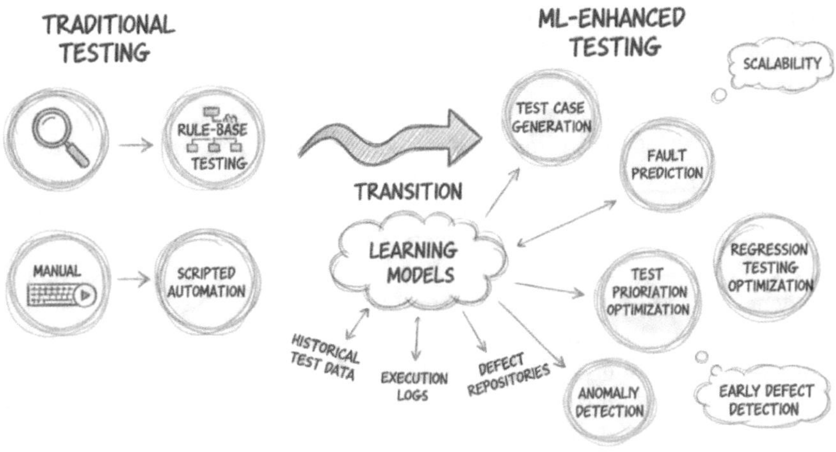

FIGURE 11.3: ML in Software Testing.

also used Support Vector Machines (SVMs), which mathematically distinguish defective modules from others. SVMs mapped software features into a higher-dimensional space to find the best category boundary, unlike tree-based models. This method improved defect classification by letting SVMs manage complex feature interactions. These models outperformed rule-based systems in defect detection only after extensive feature engineering. Manually selecting and preprocessing relevant features required software engineering and ML skills. Early successful software testing ML models relied on feature engineering. To train models effectively, domain experts had to extract valuable attributes from the codebase since these algorithms could not directly process raw code. Such features as functionality, change frequency, cyclomatic complexity, code coupling metrics, and defect density were also prevalent. These dimensions allowed for ML models to model the structural and historical characteristics of the software for defects. Cyclomatic complexity measures the number of independent path through source code used to determine program maintainability and defect risk. The code coupling measure looked at the interaction between various software modules to identify areas where changes were most likely to cause bugs. Feature engineering improved the accuracy of ML models over rule-based models but had drawbacks. Manual feature extraction, requiring domain expertise, compelled testing of software that was ML driven.

11.6 Deep Learning-Based Defect Detection

When DL became the dominant AI paradigm, scientists began to use neural networks in software QA. Unlike the classical ML models, DL models could analyze code representations and find patterns without human intervention. CNNs are Convolutional Neural Networks (CNNs) [44] and Long Short-Term Memory (LSTM) [18] Networks were the ideal defect detection tools for structural code relationships and sequential dependencies. CNNs were derived from image recognition programs to view source code in token sequences to be used to analyze software. They recognized regular bug patterns, security vulnerabilities, and structural differences better than feature-based ML models. LSTM sequential data processing networks processed digital traces, software execution logs, and dynamic behavior analysis. They captured long-range dependencies to reveal performance problems, failures unexpectedly and runtime issues. DL brought the power of AI testing to LLMs such as GPT-4o and BERT, extending SQA beyond model-based ML with feature specificity. These powerful AI models simplify software quality assurance and greatly improve today's software testing problem-identification. DL allows models to understand complicated patterns from source code without human feature engineering, transforming fault identification in software quality assurance. Traditional ML models with predefined characteristics made source code trait extraction difficult and domain-specific. However, DL models automatically uncover dependencies, trends, and anomalies in huge codebases, revolutionising software testing and this covers CNNs, RNNs, LSTMs. Figure 11.4 illustrates the deep learning-based defect detection framework integrating CNN, RNN, and LSTM models for software quality assurance.

11.7 Conclusions

ML, DL, and NLP are used in AI testing tools to enhance software quality assurance. These modern systems can assemble tests based on software specifications, analyze many millions of lines of source code and detect anomalies in real time. Instead, AI-driven methods rely on new information and are more accurate with time than rule-based methods. Automated regression analysis, self-healing test scripts and automated defect classification decrease manual testing and increase reliability. A suite of AI-based testing frameworks within CI/CD pipelines help to detect and resolve defects quickly and seamlessly. LLMs like GPT-4o, BERT, and CodeBERT have helped to reinvent the processes of software quality assurance in the areas of detecting defects and analysing code. These models assist developers with automated debugging, code reviews, and performance optimization. Natural language descriptions of software defects are also understood. Today, software testing largely relies on AI methodologies. Similarly, the contextual analysis, security vulnerability detection, and code refactoring of LLMs enhance reliability and efficiencies.

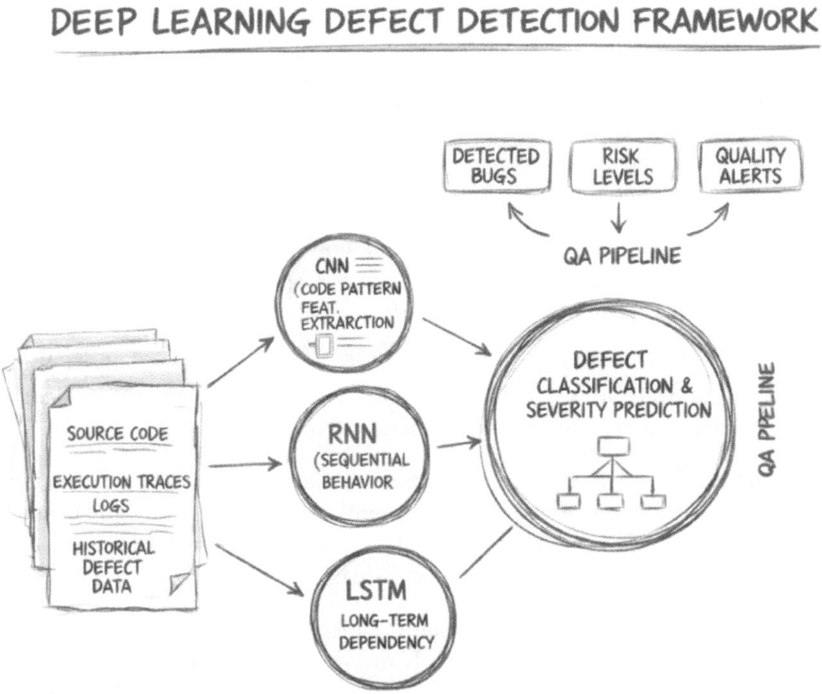

FIGURE 11.4: Deep Learning-Based Defect Detection Framework Integrating CNN, RNN, and LSTM Models for Software Quality Assurance.

Bibliography

[1] Rianat Abbas, Rasheed Afolabi, Ifeoma Eleweke, Adetomiwa Adesokan, Ahmed Akinsola, and Laticbe Elijah. Adopting secure software development practices to improve financial transactions in the banking sector. 2025.

[2] Amr Abo-eleneen, Ahammed Palliyali, and Cagatay Catal. The role of reinforcement learning in software testing. *Information and Software Technology*, 164:107325, 2023.

[3] Zohreh Aghababaeyan, Manel Abdellatif, Lionel Briand, Mojtaba Bagherzadeh, et al. Black-box testing of deep neural networks through test case diversity. *IEEE Transactions on Software Engineering*, 49(5):3182–3204, 2023.

[4] Mohammad Azzeh, Yousef Elsheikh, Ali Bou Nassif, and Lefteris Angelis. Examining the performance of kernel methods for software defect

prediction based on support vector machine. *Science of Computer Programming*, 226:102916, 2023.

[5] Aashi Singh Bhadouria. Study of: Impact of malicious attacks and data breach on the growth and performance of the company and few of the world's biggest data breaches. *International Journal of Scientific and Research Publications*, 10(10):1–11, 2022.

[6] Markus Borg. Agility in software 2.0–notebook interfaces and mlops with buttresses and rebars. In *International Conference on Lean and Agile Software Development*, pages 3–16. Springer, 2022.

[7] Marcin Borowiec, Rafał Piszko, and Tomasz Rak. Knowledge extraction and discovery about web system based on the benchmark application of online stock trading system. *Sensors*, 23(4):2274, 2023.

[8] Jürgen Börstler, Kwabena E Bennin, Sara Hooshangi, Johan Jeuring, Hieke Keuning, Carsten Kleiner, Bonnie MacKellar, Rodrigo Duran, Harald Störrle, Daniel Toll, et al. Developers talking about code quality. *Empirical Software Engineering*, 28(6):128, 2023.

[9] Hoa Khanh Dam, Trang Pham, Shien Wee Ng, Truyen Tran, John Grundy, Aditya Ghose, Taeksu Kim, and Chul-Joo Kim. Lessons learned from using a deep tree-based model for software defect prediction in practice. In *2019 IEEE/ACM 16th international conference on mining software repositories (MSR)*, pages 46–57. IEEE, 2019.

[10] Jingyu Fan. Legal policies failing on data breaches?–an empirical study of us information security law implementations. *Procedia Computer Science*, 221:971–978, 2023.

[11] Shimels D Garomssa, Rathimala Kannan, Ian Chai, and Dirk Riehle. How software quality mediates the impact of intellectual capital on commercial open-source software company success. *IEEE Access*, 10:46490–46503, 2022.

[12] Vijay Govindarajan, Pratik Patel, Sahil Tripathi, Md Azizul Hoque, and Gautam Siddharth Kashyap. Magic-enhanced keyword prompting for zero-shot audio captioning with clip models. *arXiv preprint arXiv:2509.12591*, 2025.

[13] Mark L Graber, Nancy Franklin, and Ruthanna Gordon. Diagnostic error in internal medicine. *Archives of internal medicine*, 165(13):1493–1499, 2005.

[14] Kenish Rajesh Halani, Rahul Saxena, et al. Critical analysis of manual versus automation testing. In *2021 International Conference on Computational Performance Evaluation (ComPE)*, pages 132–135. IEEE, 2021.

[15] Hongyi Han. From "flash crashes" to "gamestop": Paradigm shift in algorithm-driven market crises. *Finance & Economics*, 1(6), 2025.

[16] Xinyi Hou, Yanjie Zhao, Yue Liu, Zhou Yang, Kailong Wang, Li Li, Xiapu Luo, David Lo, John Grundy, and Haoyu Wang. Large language models for software engineering: A systematic literature review. *ACM Transactions on Software Engineering and Methodology*, 33(8):1–79, 2024.

[17] Xing Hu, Feifei Niu, Junkai Chen, Xin Zhou, Junwei Zhang, Junda He, Xin Xia, and David Lo. Assessing and advancing benchmarks for evaluating large language models in software engineering tasks. *ACM Transactions on Software Engineering and Methodology*, 2025.

[18] Junjun Li, Shi Ying, Tiangang Li, and Xiangbo Tian. Tracedae: Trace-based anomaly detection in micro-service systems via dual autoencoder. *IEEE Transactions on Network and Service Management*, 2025.

[19] Zixi Liu, Yang Feng, Yining Yin, and Zhenyu Chen. Deepstate: selecting test suites to enhance the robustness of recurrent neural networks. In *Proceedings of the 44th International Conference on Software Engineering*, pages 598–609, 2022.

[20] Ruchika Malhotra. A systematic review of machine learning techniques for software fault prediction. *Applied Soft Computing*, 27:504–518, 2015.

[21] DA Meedeniya, ID Rubasinghe, and I Perera. Software artefacts consistency management towards continuous integration: a roadmap. *International Journal of Advanced Computer Science and Applications*, 10(4), 2019.

[22] Tianjun Mo, Pengfei Li, and Ziyi Jiang. Comparative analysis of large language models' performance in identifying different types of code defects during automated code review. *Annals of Applied Sciences*, 5(1), 2024.

[23] Partha Sarathi Mohapatra. Artificial intelligence and machine learning for test engineers: concepts in software quality assurance. *Intelligent Assurance: Artificial Intelligence-Powered Software Testing in the Modern Development Lifecycle*, 4:17, 2025.

[24] Ali Bou Nassif, Manar Abu Talib, Qassim Nasir, and Fatima Mohamad Dakalbab. Machine learning for anomaly detection: A systematic review. *Ieee Access*, 9:78658–78700, 2021.

[25] Harshad Vijay Pandhare. Future of software test automation using ai/ml. *International Journal Of Engineering And Computer Science*, 13(05), 2025.

[26] George Papadimitriou, Dimitris Gizopoulos, Harish Dattatraya Dixit, and Sriram Sankar. Silent data corruptions: The stealthy saboteurs of digital integrity. In *2023 IEEE 29th International Symposium on On-Line Testing and Robust System Design (IOLTS)*, pages 1–7. IEEE, 2023.

[27] Sandeep Reddy. Global harmonization of artificial intelligence-enabled software as a medical device regulation: addressing challenges and unifying standards. *Mayo Clinic Proceedings: Digital Health*, 3(1), 2025.

[28] Elisabetta Ronchieri and Marco Canaparo. Assessing the impact of software quality models in healthcare software systems. *Health Systems*, 12(1):85–97, 2023.

[29] Daniel Russo, Paolo Ciancarini, Tommaso Falasconi, and Massimo Tomasi. Software quality concerns in the italian bank sector: the emergence of a meta-quality dimension. In *2017 IEEE/ACM 39th International Conference on Software Engineering: Software Engineering in Practice Track (ICSE-SEIP)*, pages 63–72. IEEE, 2017.

[30] Hozyfa Shafa and Mst Shahrin Sultana. The influence of secure data systems on fraud detection in business intelligence applications. *Journal of Sustainable Development and Policy*, 3(04):133–173, 2024.

[31] Vikram Singh, Srishti Kapoor, and Lavanya Jha. Quality assurance through automated testing frameworks in software engineering. *Journal of Software Engineering & Software Testing ISSN: 2457-0516 (Online)*, 10(2), 2025.

[32] Arpita Soni, Sahil Tripathi, Gautam Siddharth Kashyap, Manaswi Kulahara, Mohammad Anas Azeez, Zohaib Hasan Siddiqui, Nipun Joshi, and Jiechao Gao. Can we predict your next move without breaking your privacy? *arXiv preprint arXiv:2507.08843*, 2025.

[33] Inês Sousa Sousa, António Casimiro, and José Cecílio. Artificial neural networks for real-time data quality assurance. *ACM SIGAda Ada Letters*, 42(1):86–89, 2022.

[34] Reudismam Sousa, Gustavo Soares, Rohit Gheyi, Titus Barik, and Loris D'Antoni. Learning quick fixes from code repositories. In *Proceedings of the XXXV Brazilian Symposium on Software Engineering*, pages 74–83, 2021.

[35] Noone Srinivas, Nagaraj Mandaloju, and Siddhartha Varma Nadimpalli. Leveraging automation in software quality assurance: Enhancing defect detection and improving efficiency. *International Journal of Acta Informatica*, 3(1):112–124, 2024.

[36] Swetha Talakola. The optimization of software testing efficiency and effectiveness using ai techniques. *International Journal of Artificial Intelligence, Data Science, and Machine Learning*, 5(3):23–34, 2024.

[37] Florian Tambon, Arghavan Moradi-Dakhel, Amin Nikanjam, Foutse Khomh, Michel C Desmarais, and Giuliano Antoniol. Bugs in large language models generated code: An empirical study. *Empirical Software Engineering*, 30(3):65, 2025.

[38] Yu TANG, Qi DAI, Zhi-wei YANG, Ai-min YANG, and Li-fang CHEN. A software defect prediction algorithm based on optimized random forest. *Computer Engineering & Science*, 45(05):830, 2023.

[39] Cansu Tayaksi, Erhan Ada, Yigit Kazancoglu, and Muhittin Sagnak. The financial impacts of information systems security breaches on publicly traded companies: reactions of different sectors. *Journal of Enterprise Information Management*, 35(2):650–668, 2022.

[40] Junjie Wang, Yuchao Huang, Chunyang Chen, Zhe Liu, Song Wang, and Qing Wang. Software testing with large language models: Survey, landscape, and vision. *IEEE Transactions on Software Engineering*, 50(4):911–936, 2024.

[41] Zhilong Wang, Chen Cao, Li Yu, Suhang Wang, Xiaoyan Sun, and Peng Liu. Deepsanitizer: Combining heuristic rules and deep learning models to spot silent buffer overflows in binary. *IEEE Transactions on Dependable and Secure Computing*, 2025.

[42] Jones Yeboah and Saheed Popoola. Analyzing user sentiments towards static analysis tools: A study using user reviews. In *2024 2nd International Conference on Artificial Intelligence, Blockchain, and Internet of Things (AIBThings)*, pages 1–7. IEEE, 2024.

[43] Quanjun Zhang, Chunrong Fang, Yang Xie, YuXiang Ma, Weisong Sun, Yun Yang, and Zhenyu Chen. A systematic literature review on large language models for automated program repair. *arXiv preprint arXiv:2405.01466*, 2024.

[44] Yajie Zhang. Software defect prediction model based on deep learning. In *2024 International Conference on Power, Electrical Engineering, Electronics and Control (PEEEC)*, pages 954–959. IEEE, 2024.

[45] Qiang Zhi, Li Gong, Jianguo Ren, Meiyu Liu, Zhengshu Zhou, and Shuichiro Yamamoto. Element quality indicator: A quality assessment and defect detection method for software requirement specification. *Heliyon*, 9(5), 2023.

[46] Mariusz Zieja, Andrzej Szelmanowski, Andrzej Pazur, and Grzegorz Kowalczyk. Computer life-cycle management system for avionics software as a tool for supporting the sustainable development of air transport. *Sustainability*, 13(3):1547, 2021.

Chapter 12

Enhancing Software Reliability Prediction with Machine Learning: Addressing Data Noise and Model Consensus

John Akkarakaran Jose

12.1 Introduction

ML is seen as an attractive means of addressing the issues related to the reliability of software used [6]. But there are a number of challenges that need to be overcome in order to tap the potential of this. First, the scope of the uses of ML is limited by the absence of publicly accessible and applicable data. The procedure is very complicated as such and gathering quality and representative data is not an easy task. Secondly, the experimentally observed data of software reliability typically contain inevitable inaccuracies with ± 0.5- 0.7 LogS. Such errors may affect the quality of the ML models, so it is necessary to be attentive on handling data and handling the errors. Lastly, it is not an easy job to choose the right models and methods of software reliability. Nevertheless, these challenges have not eliminated the possibility of progressing towards these obstacles due to the continual development of computational techniques and the increased availability of ML tools. Software reliability would transform the drug development process into something accurate, and enable the researcher to optimize the design process and allocate resources to compounds that have the most desirable qualities through software reliability [3]. The growing interest in dealing with such challenges suggests that the issue of software reliability, although complicated, is not an impossible task. The integration of ML into this field is still ongoing and continues to enrich the field and predict the solubility with more precision [10]. Hence, the solubility is one of the cornerstone properties, which determines various areas of drug development, crystallization, and the chemical processing. It is essential in realizing the efficacy and safety of drugs, a fact that is reflected on the critical role

DOI: 10.1201/9781003774815-12

TABLE 12.1: Survey of Machine Learning Approaches in Software Reliability Modeling

Aspect	Technique	Outcome
Descriptors	Physicochemical	Correlation
Datasets	Curated	Availability
Modeling	Regression	Baseline
Learning	Neural	Accuracy
Ensembles	Combination	Stability
Optimization	Tuning	Performance
Validation	Crossvalidation	Reliability
Complexity	Nonlinear	Flexibility
Prediction	Forests	Robustness
Generalization	Regularization	Consistency
Coverage	Chemicalspace	Representativeness
Evaluation	Benchmarking	Comparison
Scalability	Automation	Efficiency

it plays in determining ADMET properties. Additionally, software reliability may greatly decrease the expenses and work force involved in designing drugs and compounds.

The chapter goes as follows; the corresponding works will be presented in the Section 12.3. It describes the materials and methods having experimental analysis which is discussed in Section 12.4. We analyze the results in Section 12.5 and conclude the chapter with a few suggestions on how the research could be furthered in Section 12.6.

12.2 Related Works

Software reliability has been highly developed (see Table 12.1) with initial studies aiming at determining molecular characteristics that define solubility like the octanol-water partition coefficient, the melting point, and the area of the molecules. These findings provided the basis of solubility and molecular descriptors relations. With the introduction of ML, however, the situation changed, and more complex models became possible to deal with the complexity of software reliability. Regardless of the improvements, data quality and availability are major issues. The presentation of major datasets, including the Aquasol catalyzed progress in the form of the underlining data of ML models. Creative work such as Huuskonen dataset and Delaney dataset were also useful in training models and testing. With the introduction of curated resources such as AqSolDB, there was an expansion of data availability, which is absent in the past, and the accuracy of predicting data is improved. The modeling strategies have developed together with the growth of data. Early software reliabilitys were based on linear regression model with the General Solubility Equation (GSE) as a breakthrough. But with the increase in the computational power, more sophisticated algorithms such as the Artificial Neural Networks (ANN) [2], random forests, and Principal Component Regression (PCR) proved better. As an example, ANN was found to work better than linear regression, whereas PCR models obtained the same or even better results. Events such as the solubility challenge brought out the need to have true data and model strength. The challenges remain, though, such as experimental errors, variability of data and extensive coverage of chemical-space is required. More recent approaches, including ensemble modeling [356-358] and new algorithms, like SCRATCH [20], aim to address such concerns through the use of a wide range of data and prediction methods. The development of software reliability models has been characterized by the gradual growth in the use of data and the use of ML methods. Initial investigations, including those of, have utilized Recurring Neural Networks (RNN) with mixed datasets and have attained good R-squared scores, up to 0.95, which shows the promise of neural networks. Equally, established that neural networks surpassed the classical Multiple Linear Regression (MLR). Cubist models which were introduced with better performance than Partial Least Squares (PLS) models, and used Support Vector Regression (SVR) with sufficient accuracy. The change of focus occurred as the studies started to focus more on the significance of relevant descriptors rather than on the volume of data. Obtained strong results using only nine molecular descriptors, which illustrates the quality of data. Random forest models started to popularize in 2007 as they showed better performance than SVR and ANNs, whereas Gaussian processes also showed high predictive performance. The publication of the solubility challenge dataset in 2008 and other works highlighted the necessity to have more quality-based data and models that are stronger and more effective in quality. Ensemble techniques became popular, including, where multiple models are used to make predictions and the stability and accuracy were enhanced by the combination of the predictions.

12.3 Materials and Methods

This section gives a summary of the software reliability methodology giving the outline of the process and then proceeds to explain each of them in the following sections. Figure 12.1 summarises the flow of the operations, which shows how data is processed, models are generated and tested, and predictions are being perfected to bring results which are desired. The mentioned methodology is focused on the precision of the step-by-step solution and the need to have a solid background on the complexities in the future. The data acquisition and preparation are the first step. The data is purged to overcome null values, outliers and discrepancies so as to produce quality inputs to the

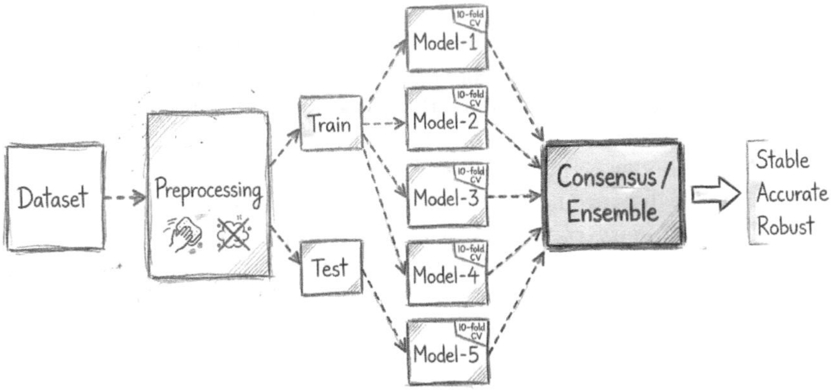

FIGURE 12.1: Architecture Providing an Overview of the Proposed Methodology.

modeling. The next step involves splitting the data into training and testing sets after this preprocessing step to model development and evaluation. The training set is used to build several ML models, which in this case, due to simplicity, are represented by five models (Model-1 through Model-5). All the models are evaluated with a strict series of 10-fold cross-validation. This is a critical step because it guarantees the quality of performance of the model is not overfitted to the training set. The results of cross-validation give the necessary metrics of the performance, which are archived to be compared later. Besides set evaluation, the performance of each model is also evaluated and the performance results on the test set are captured. These measures allow to comprehensively compare the models and determine the best performers predator and predictor. After establishing the models that work best, one then proceeds to the ensemble stage, commonly known as the wisdom of the crowd. The predictions on the selected models are combined to produce a consensus prediction on the test set. This group approach takes advantage of the strengths of several models and dilutes the weaknesses of each and enhances the general predictive accuracy. Lastly, an evaluation of the effectiveness of the aggregated prediction is done by measuring the performance with the help of established evaluation metrics, to have a comprehensive picture of how effective the methodology is.

12.3.1 Dataset Analysis

The data preparation before modeling is a crucial process towards making accurate and reliable predictions as suggested by [1]. The main issues are working with missing values, outliers and highly correlated independent variables which, without effective solutions, may cause noise or model distortion.

The absent values should be handled with care to prevent the extraneous noise. Removing records with missing values is an easy solution but in a smaller set, it might be impossible to lose the information contained in a record. At this, imputation techniques, e.g., substituting all missing values with the mean, median, or predictions made by a ML algorithm can be used. A major consideration should also be given to outliers, which are information points that differ considerably as compared to the norm. Removing them is possible, but more rational values would be better to input in situations where data integrity is the primary concern. Correlated features that are very high also pose a challenge, since they have the power to boost regression coefficients and make the model less reliable. To overcome this, it is possible to remove one of the features that are highly correlated without much information loss. The correlation coefficient between two continuous variables is defined as the Pearson correlation coefficient (r) of variables commonly used to determine such correlations. The correlation coefficient is also an important measure because it is used to assess the linear relationship between two variables, which takes values between –1 and +1. Values close to zero signify weak or no relation whereas the values nearer to –1 or +1 signify stronger negative or positive linear relations respectively. A positive r will mean a positive relationship where a negative r means an inverse relationship. When discussing our chapter on chemical compounds, a squared correlation value (r-squared) of a value greater than 0.9 is highly correlated according to the threshold that was set by [9]. This means that variables whose r-squared is more than 0.9 would include almost the same information, making either of them redundant. In a model that has one or several independent variables whose correlation coefficients are high, regression coefficients may be inflated, causing unreliable predictions- multicollinearity. To solve this, the Variance Inflation Factor (VIF) is applied to measure the level of multicollinearity. VIF is used to determine the degree of increasing the variance of regression coefficients because of the relationship between the variables. In simple terms, it can be used to measure the extent to which multicollinearity affects the stability of the coefficients. VIF computation is based on Coefficient of Determination (R^2), which is one of the basic measures of quality of regression model.

In this analysis, the data will be divided into 80% to be used in the training process and 20% to be used in the testing process. The models will be built to use the training data and assess the performance of the models by comparing them with cross-validation. Cross-validation will produce forecasts on the training set and the generalization performance of the model will be determined. These test set will be evaluated using a final evaluation of model performance, which will be done using the test set that will be maintained as a separate set.

12.3.2 Model Analysis

The "wisdom of the crowd" theory emphasizes that the aggregates of estimations of various people could provide more precise and valid results than personal estimates. The idea is consistent with the jury theory developed by Condorcet which showed that the likelihood of correct choice in majority votes is growing with the number of independent and competent judges on the jury board [5]. This principle is used in ML to ensemble learning, in which different and independently executing models are combined to improve the accuracy of prediction. This is referred to as consensus prediction which minimizes model complexity, overfitting and provides more stable results. A number of ML techniques are used in constructing ensemble methods. Ordinary Least Squares (OLS) regression [4], also referred to as the modern or Neoclassical regression, is used to estimate a linear relationship between independent and dependent variables by minimizing the total squared error. OSL however is susceptible to multicollinearity and overfitting. Building on the use of OLS, Ridge regression eliminates large coefficient values by including a regularization term to shrink the coefficients to minimize complexity and overfitting [7]. Likewise, lasso regression [8] which is another regression method, not only contracts the coefficients but also pushes some to zero, thus feature selection. The SVR uses support vectors to build a street around data points making outliers to have little influence and concentrate more on the data within a defined margin of tolerance (ϵ) .

Its loss functional that is epsilon-insensitive makes sure that the prediction is not affected by small changes in the margin, which makes the models robust. PCR and PLS dimensionality reduction methods are additional ways of enhancing model performance. The independent variables are converted into uncorrelated principal components by PCR which get the maximum variance. These elements make the regression process complex simple because they minimize dimensionality and multicollinearity. Instead, PLS places value on variance in the dependent variable in constructing components which increases the predictive strength where the independent variables have high correlation with the target. Ensemble methods build on such a great variety of approaches to employ collective intelligence.

12.4 Experimental Analysis

12.4.1 Experimental Setup

The chapter primarily relies on two powerful libraries: scikit-learn, a widely-used package for efficient ML, and GPy, a robust framework for Gaussian process modeling. The OLS method, implemented via `sklearn.linear_model.LinearRegression`, is the foundation for linear regression. Ridge Regression (RR) and Lasso Regression (LaR), regularized extensions of OLS, are handled using `sklearn.linear_model.Ridge` and `sklearn.linear_model.Lasso`, respectively. SVR, focusing on maximizing margins, is implemented with `sklearn.svm.LinearSVR`. For dimensionality reduction techniques, PCR utilizes `sklearn.decomposition.PCA`, while PLS is implemented through `sklearn.cross_decomposition.PLSRegression`. Bagging methods using decision trees are managed via `sklearn.ensemble.BaggingRegressor`, and random forests are implemented with `sklearn.ensemble. RandomForestRegressor`. Extremely Randomized Trees (ERT), offering a variation on decision-tree ensembles, are accessed through `sklearn.ensemble. ExtraTreesRegressor`. Boosting methods, including Adaptive Boosting (ADAB) and Gradient Boosting (GRADB), are implemented with `sklearn.ensemble.AdaBoostRegressor` and `sklearn.ensemble.GradientBoostingRegressor`, respectively. Non-linear Support Vector Regression (KSVR) uses `sklearn.svm.SVR`. Gaussian process regression is supported by the GPy library (`GPy.models.GPRegression`), providing a Bayesian approach to regression. Lastly, neural network-based regression through Multi-layer Perceptrons (MLP) is handled with `sklearn.neural_network.MLPRegressor`. This suite of algorithms and tools demonstrates the versatility and strength of scikit-learn and GPy in implementing advanced ML workflows.

12.4.2 Evaluation Metrics

Testing the predictions of a model is important when it is necessary to determine its reliability and to make improvements. Here, we dwell on solubility prediction on a variety of evaluation measurements. Comparing predicted and actual solubility values, we gain insights into model performance. Key metrics include Root Mean Square Error (RMSE), Pearson's correlation coefficient squared (r^2) and extra accuracy metrics like %P$\pm$0.7 and %P$\pm$1.0 which is used to solve experimental variability. RMSE measures the average difference between predicted and actual values which is been calculated as the square root of the mean squared residuals. Its value ranges from 0 to infinity, with lower values indicating better performance. Widely used in regression analysis, RMSE provides a clear measure of prediction accuracy and model fit. Pearson's r^2 has derived from the correlation coefficient, quantifies the proportion of variance explained by the model. Ranging from 0 to 1 it is particularly useful for comparing results with prior studies in software reliability. To solve for experimental variability %P$\pm$0.7 and %P$\pm$1.0 metrics evaluate the percentage of predictions within $\pm$0.7 and $\pm$1.0 intervals of experimental values. These metrics act as measures of prediction accuracy with higher percentages showing better model performance. The chapter shows %P$\pm$0.7 as the primary metric given its alignment with expected experimental error which is been followed by %P$\pm$1.0, r^2 and RMSE. This systematic evaluation framework shows a good understanding of model predictions by focusing accuracy and performance.

TABLE 12.2: Performance of Ordinary Least Squares (OLS) Regression

Metric	RMSE	R^2
Cross-Validation	1.25	0.42
Test Set	1.38	0.36
$\%P \pm 0.7$	48%	–

TABLE 12.3: Residual Analysis of OLS Regression Model

Metric	Value	Interpretation
Mean Residual	0.12	Biased
Residual Variance	1.54	High
Pattern Presence	Yes	Non-random

12.5 Result Analysis

12.5.1 Ordinary Linear Regression Models

The analysis of the different linear models used in predicting the solubility indicated that similar difficulties were encountered which are mainly attributed to the fact that there is no linear association between the independent and the dependent variables. The OLS model was constructed using default hyperparameters and the predictions exhibited a lot of scatter as shown in Table 12.2. Although OLS slightly outperformed on cross-validation as compared to the test set, such a variance indicates overfitting. The overall performance as the percent of P with a 95% confidence interval of 0.7 was less than 50% indicating that the number of predictions in the acceptable range was less than half. The residual-predictions (see Table 12.3) also corroborated the fact that the residual was not distributed in an accidental fashion, which implied that a linear model should not be applicable to the problem. Further issues were revealed in a closer examination of the coefficients of the model as well as the correlation between the variables. Multicollinearity issues were encountered with the variables of Molecular Weight (MV) and Solvent-Accessible Surface Area (SASA), which had strong correlations with each other, as these variables gave coefficients that were not reliable. The deletion of SASA enhanced the meaning and relevance of MW, however this did not answer the inherent constraints of the linear models. Although such problems are evident, SASA was not phased out because it is relevant to the reliability of software.

The Ridge Regression model with tuned hyperparameter has been showed minor improvements over OLS due to regularization as shown in Table 12.4.

TABLE 12.4: Performance of Ridge Regression Model

Metric	RMSE	R^2
Cross-Validation	1.18	0.45
Test Set	1.32	0.39
$\%P \pm 0.7$	50%	–

TABLE 12.5: Performance of Lasso Regression Model

Metric	RMSE	R^2
Cross-Validation	1.15	0.47
Test Set	1.29	0.38
$\%P \pm 0.7$	49%	–

TABLE 12.6: Performance of Linear Support Vector Regression (SVR)

Metric	RMSE	R^2
Cross-Validation	1.22	0.41
Test Set	1.20	0.44
$\%P \pm 0.7$	52%	–

This included a slight increase in %P±0.7 by reaching exactly 50% and some shrinkage in coefficient values. The residual-predictions has confirmed that the lack of linearity is going to remain a fundamental barrier by limiting performance gains. Lasso Regression was also applied providing the best results as shown in Table 12.5. While Lasso effectively reduced many coefficients to zero, the overall improvement in prediction accuracy was minimal. %P±0.7 improved slightly in the CV set, but performance on the test set declined. Although the RMSE on the test set was marginally better than OLS and Ridge, the lack of consistent improvement across metrics indicated that Lasso could not overcome the fundamental nonlinearity of the data. Linear SVR produced results similar to the previous linear models as shown in Table 12.6. Interestingly, SVR performed better on the test set than in cross-validation, achieving higher %P±0.7 scores for the test data. However, other metrics, such as RMSE and r^2 showed little improvement over OLS, Ridge, or Lasso. The persistent lack of a linear relationship between variables, as highlighted in residual-prediction plots, explains the limited performance of SVR despite its flexibility compared to OLS.

TABLE 12.7: Explained Variance of Principal Components

Component Range	Explained Variance	Cumulative Variance
PC1–PC3	68.4	68.4
PC1–PC5	86.7	86.7
PC1–PC7	95.6	95.6

TABLE 12.8: Selection of Principal Components Based on $\%P \pm 0.7$

Number of PCs	$\%P \pm 0.7$ (CV)	$\%P \pm 0.7$ (Test)
7 PCs	52%	50%
9 PCs	56%	54%
10 PCs	60%	59%

TABLE 12.9: Performance of Principal Component Regression Model

Metric	RMSE	R^2
Cross-Validation	1.08	0.52
Test Set	1.15	0.48
$\%P \pm 0.7$	60%	–

TABLE 12.10: Stability Analysis of PCR Model Performance

Dataset	$\%P \pm 0.7$	RMSE
Cross-Validation	60%	1.08
Test Set	59%	1.15
Difference	1%	0.07

12.5.2 Latent Variable Models

PCA was applied to the training data, yielding eleven principal components corresponding to the eleven independent variables (descriptors). As shown in Table 12.7, the first seven components accounted for over 95% of the variance in the data. However, the selection of components should not solely depend on the explained variance of independent variables. Low-variance components can still have significant correlations with the dependent variable, and discarding them might reduce the predictive power of the model. To address this, models with varying numbers of components were evaluated using cross-validation, leading to the selection of the first ten components as optimal based on %P±0.7 results, as shown in Table 12.8.

A linear regression model was then fitted using these ten components. The performance, as indicated in Tables 12.9 and 12.10, showed some improvement compared to previous linear models, particularly in terms of %P±0.7.

TABLE 12.11: Performance of Partial Least Squares (PLS) Regression Model

Number of Components	RMSE	R^2
5 Components	1.12	0.49
7 Components	1.04	0.55
9 Components	1.06	0.53

TABLE 12.12: Performance of Bagging Ensemble Model

Metric	RMSE	R^2
Cross-Validation	0.95	0.65
Test Set	1.02	0.61
$\%P \pm 0.7$	68%	–

The stability of the %P±0.7 metric across cross-validation and test sets has suggested improved model consistency. This can be attributed to the orthogonality of principal components which will reduce multicollinearity among predictors. Despite these improvements the overall performance did not show good enhancement which further supports the conclusion that the lack of a linear relationship between the independent and dependent variables limits the predictive capability of linear models. PLS regression was also explored with models evaluated with the help of increasing numbers of components to determine the optimal set. Based on Table 12.11 the first seven components which were selected for the final PLS model.

12.5.3 Ensemble Learning Models

We implemented bagging, random forest, Extremely Randomised Trees (ET) and boosting algorithms to increase prediction accuracy. For bagging we have tuned the number of decision trees ($n_{\text{estimators}}$) and determined the optimal value as 500. The model has showed a good amount of improvement with a 10% increase in %P±0.7 compared to linear models. This result shows the non-linear information in the dataset which is been accessed by bagging. Although the test-set predictions were slightly less accurate than cross-validation results the model has showed stable performance overall as shown in Table 12.12.

Random forest was a variant of bagging method that increased the amount of randomness by performing random feature selection and depth limitation. We used hyperparameters optimization with $n_{\text{estimators}} = 100$, max features $=$ 7, and max depth $= 50$. Random forest marginally improved the performance of bagging with respect to the test set, by having a lower variance because of the random selection of features as indicated in Table 12.13. The two features,

TABLE 12.13:
Performance of Random Forest Regression Model

Metric	**RMSE**	R^2
Cross-Validation	0.90	0.69
Test Set	0.96	0.66
$\%P \pm 0.7$	71%	–

TABLE 12.14:
Performance of Extremely Randomized Trees (ET) Model

Metric	**RMSE**	R^2
Cross-Validation	0.82	0.75
Test Set	0.88	0.72
$\%P \pm 0.7$	76%	–

TABLE 12.15:
Performance of AdaBoost Regression Model

Metric	**RMSE**	R^2
Cross-Validation	1.10	0.52
Test Set	1.18	0.48
$\%P \pm 0.7$	60%	–

namely, SASA and MW, turned out to be the most popular, adding to their importance. The improved performance of RF is one of the examples that can demonstrate the benefits of controlled randomness in ensemble learning. ET was found to be the most performing bagging strategy. Using 100 decision trees, the model had more predictions concentrated around the line of best fit by evaluating metrics of evaluation and feature importance in Table 12.14. The most influential variable was found to be 'MW', then it was succeeded by the influence of the variables of SASA and Most neg. The additional randomness in ET further minimized variance and provided better results than bagging and RF. The tuning of ADAB in the boosting category was done using n estimators = 1000, and max depth = 5. ADAB did not perform better than bagging models as depicted in Table 12.15 because it failed to bring meaningful information as compared to bagging models. This feature of the algorithm of re-weighting the misclassified samples could occasionally create difficulties in reducing the losses, which restricts its capability to perform the optimal learning.

TABLE 12.16: Performance of Kernel Support Vector Regression (KSVR) on Cross-Validation

Metric	Mean Absolute Error	R^2
Training Error	0.82	0.72
Validation Error	0.88	0.69
Overall Performance	0.85	0.71

TABLE 12.17: Performance of Kernel Support Vector Regression (KSVR) on Test Data

Metric	Mean Absolute Error	R^2
Training Set	0.80	0.74
Test Set	0.90	0.68
Overall Performance	0.86	0.71

12.5.4 Kernel-Based Models

Performance of KSVR and Gaussian process regression in predicting solubility, focusing on the advantages of kernel-based methods in modeling non-linear relationships. For KSVR, key hyperparameters—epsilon (ϵ), regularization coefficient (C), and the RBF kernel coefficient (γ)—were tuned to optimal values: $\epsilon = 0.069$, $C = 5$, and $\gamma = 0.0909$. Unlike linear models, KSVR transforms input features into a high-dimensional space using kernel transformations, enabling the model to capture complex, non-linear patterns. However, this transformation makes it infeasible to directly assess feature importance. KSVR outperformed its linear counterpart, demonstrating that kernel transformations significantly enhance the model's ability to learn intricate patterns inaccessible to linear models. Performance metrics presented in Tables 12.16 and 12.17, indicating that KSVR achieved better results than most models analyzed, reinforcing the assumption that linear models are unsuitable for software reliability tasks.

Gaussian process regression was also explored using the GPy module, which optimizes hyperparameters through a log-likelihood maximization approach. The optimally tuned parameters were RBF variance = 15.1466, RBF lengthscale = 3.780, and Gaussian noise variance = 0.6335. These parameters were further refined using cross-validation for each training fold, ensuring their robustness. The same optimized values were employed for predictions on the test set. Similar to KSVR, GPR relies on kernel transformations to represent input features in high-dimensional spaces, making feature importance retrieval infeasible. GPR's predictions, displayed in Table 12.18, demonstrated strong performance, exceeding that of all linear models and delivering accurate predictions. The Bayesian nature of GPR contributed to its success, as it is well-suited for scenarios with limited data. The algorithm leverages kernels to

TABLE 12.18: Performance of Gaussian Process Regression (GPR) Model

Metric	Mean Absolute Error	R^2
Cross-Validation	0.78	0.76
Test Set	0.85	0.73
Overall Performance	0.82	0.75

learn complex patterns, allowing it to effectively handle data with non-linear relationships.

12.6 Conclusions

The chapter brings out some of the findings in the field of software reliability. It is important to choose the right evaluation measures, including the one like %P±0.7 and %P±1.0 which should consider the errors in the experiment and enhance the confidence in the work of the model. Regularized models and other linear models performed poorly because the relationships in the data were non-linear, and regularized models, such as Ridge and Lasso Regression, slightly lowered variance. PCR was superior to PLS because it conserved well on independent variable variance. Ensemble techniques, especially ET, were found to be better in terms of more randomness and less variance. Gradient boosting was better than adaptive boosting because it has a differentiable loss and gradient descent optimization. There was a strong ability to capture non-linear patterns using kernel-based methods and make strong predictions. The benefits of collective intelligence were proven by the consensus of the best performing models. Nevertheless, the performance of the chapter was restricted due to the availability of only data, which implies that further and more qualitative data might be even more accurate.

Bibliography

[1] Faruk Bayram, Bestoun S Ahmed, and Erik Hallin. End-to-end data quality-driven framework for machine learning in production environment. *Heliyon*, 12(1), 2026.

[2] S. Falsafi and H. Karimi. A comparative study of artificial neural networks, random forest, and support vector regression for predicting aqueous solubility. *Journal of Molecular Liquids*, 395:123891, 2024.

[3] Amin Ghaedi, Akram Khatibi Bardsiri, and Mahmoud Jafari Shahbazzadeh. Software failure prediction based on game theory and convolutional neural network optimized by cat hunting optimization (cho) algorithm. *Management Strategies and Engineering Sciences*, 7(1):34–55, 2025.

[4] M. Gonzalez and R. Fernandez. Modern statistical learning: Evaluating ols limitations in the era of big data. *Journal of Computational Statistics*, 39(4):1580–1602, 2024.

[5] A. Muller and J. Schmidt. Beyond the jury: Modern applications of condorcet's theorem in decentralized ai ensembles. *Artificial Intelligence Review*, 63(2):112–134, 2025.

[6] Upasana Samal and Abhinav Kumar. A neural network approach for software reliability prediction. *International Journal of Reliability, Quality and Safety Engineering*, 31(03), 2024.

[7] K. Tanaka and H. Sato. Adaptive ridge regression for coefficient stability in high-variance predictive models. *Expert Systems with Applications*, 268:125401, 2025.

[8] L. Zhao and D. Kim. Sparsity-driven feature selection: A comprehensive review of lasso and elastic net in bioinformatics. *Digital Discovery*, 3(8):945–962, 2024.

[9] Y. Zhao and K. Miller. Feature redundancy and correlation thresholds in high-dimensional predictive modeling. *Digital Discovery*, 4(3):215–230, 2025.

[10] Tian Zheng, John BO Mitchell, and Simon Dobson. Revisiting the application of machine learning approaches in predicting aqueous solubility. *ACS Omega*, 9(31):35209–35222, 2024.

Chapter 13

AI for Cyber-Physical Security in Industrial IoT Systems

Sushant Kr. Ray

13.1 Introduction

The rapid growth of Industrial Internet of Things (IIoT) technologies have been transformed modern manufacturing [11], energy systems [13] and important infrastructure [12] by enabling real-time monitoring, automation and intelligent control. The increasing interconnection between cyber [21] and physical [8] components have also expanded the attack surface, exposing industrial systems to complex cyber-physical threats [16]. Attacks targeting sensors, actuators, communication networks and control software can disrupt operations, compromise safety and cause important economic losses.

There are some traditional security mechanisms which mainly trusts on static rules [17] and perimeter-based defenses [12] are inadequate for detecting advanced persistent threats and coordinated cyber-physical attacks [4]. Artificial Intelligence (AI) has grown as a powerful tool for increasing situational awareness [21], anomaly detection [14] and adaptive defense in IIoT environments [6]. Through learning complex behavioral patterns from large-scale data AI systems which can identify subtle deviations indicative of malicious type of activities. Figure 13.1 shows an overview of AI-enabled cyber-physical security in industrial IoT systems.

This chapter examines the role of AI in securing IIoT infrastructures in Section 13.2. Section 13.3 discusses intelligent threat detection techniques while Section 13.4 focuses on deployment challenges and real-world applications. The chapter concludes in Section 13.5 with future research directions.

DOI: 10.1201/9781003774815-13

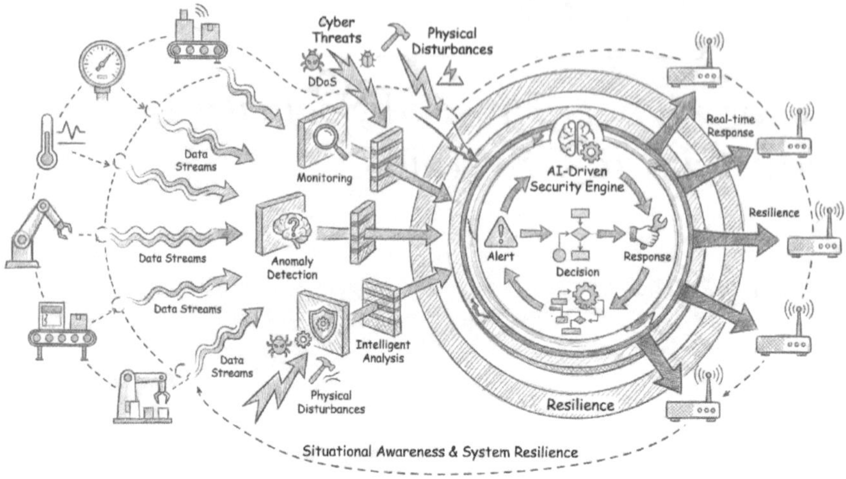

FIGURE 13.1: Overview of AI-Enabled Cyber-Physical Security in Industrial IoT Systems.

13.2 Cyber-Physical Modeling of Industrial IoT Systems

Industrial IoT systems uses physical processes with computational and networking components by forming complex cyber-physical environments [21]. Some good security solutions require accurate modeling of both physical dynamics and cyber interactions. Graph-based representations [19], digital twins [5] and hybrid system [9] models are commonly used to capture relationships among sensors, controllers, machines, and communication links.

There are some data which has collected from Programmable Logic Controllers (PLCs), supervisory control and data acquisition (SCADA) systems and edge devices is transformed into structured representations for AI analysis [15]. Temporal modeling techniques further capture system evolution and operational type of dependencies over time [10]. Feature extraction processes encode physical measurements, control commands and network traffic patterns. Figure 13.2 shows the transformation of raw IIoT data into cyber-physical models for security analytics.

13.3 AI-Based Threat Detection and Response Mechanisms

AI-driven security frameworks uses machine learning and deep learning techniques to detect, classify and respond to cyber-physical threats in IIoT

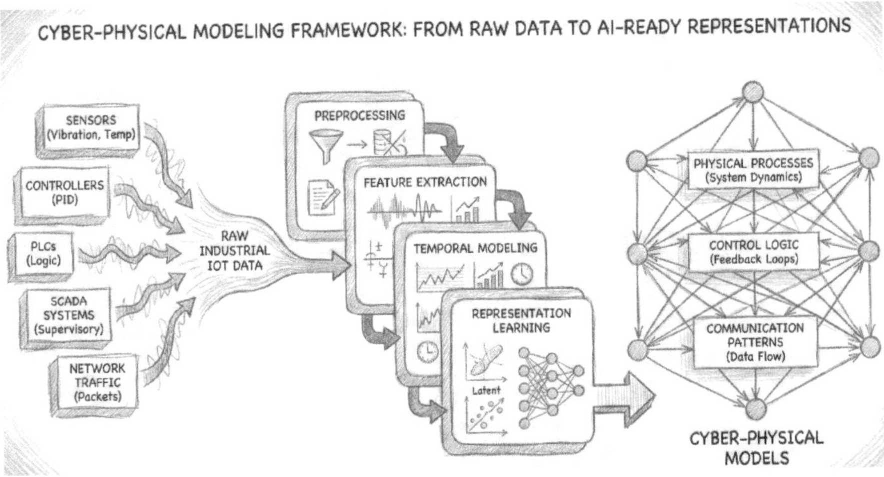

FIGURE 13.2: Cyber-Physical Modeling Framework for Transforming Raw IIoT Data into Structured Representations.

systems [17]. Unsupervised learning models like autoencoders [18] and clustering algorithms [3] which identify anomalous system behaviors without requiring any labeled attack data. Supervised classifiers and ensemble models support attack categorization and risk evaluation [1].

The architectures of deep learning (convolutional and recurrence neural networks) can process multivariate sensor streams and network traffic to detect intrusion attempts [7]. Reinforcement learning is used to learn optimal response policies in dynamic attack conditions so that adaptive defense strategies can be formed [6]. Hybrid methods involve the integration of statistical anomaly detection and semantic analysis to enhance the accuracy of detection [2]. It uses explainable AI methods to increase transparency and the trust of the operators more often and more often [20]. Figure 13.3 above shows an AI-based threat detection and response layered architecture in IIoT environments.

13.4 Deployment Challenges, Applications, and Future Directions

Deploying AI-enabled security solutions in industrial environments involves several technical and organizational challenges [17]. These include data heterogeneity [10], real-time processing constraints [3], limited computational resources at edge nodes [21], and strict reliability requirements [20]. Interoperability with legacy systems and compliance with industrial standards

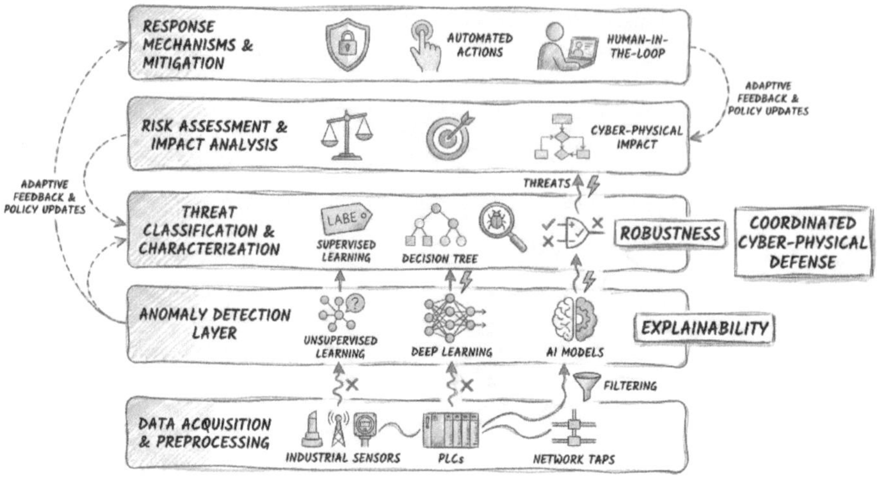

FIGURE 13.3: Layered Architecture of AI-Based Threat Detection and Response in IIoT Environments.

further complicate implementation [12]. To address these challenges, modern frameworks adopt edge-cloud collaboration [21], federated learning [2], and distributed analytics architectures [12]. Secure data sharing [20], model compression [3], and continual learning [6] mechanisms enhance scalability and adaptability. Figure 13.4 illustrates the end-to-end deployment framework of AI-driven cyber-physical security systems.

13.5 Conclusion and Future Works

This chapter has examined the use of AI in improving cyber-physical security of IIoT systems. By combining sophisticated modeling methodology, intelligent threat detection systems, and dynamic response methods, AI-based systems provide major enhancements over the traditional security models. Although these progresses have been made, there still are a number of research challenges. The future of this field should be building a robust algorithm that is resistant to adversarial influence, enhancing a better understanding of the algorithms to make them safer, and also facilitating trust between the industrial ecosystems. The privacy-preserving learning and digital twin integration and standardized benchmarking platforms are also significant directions.

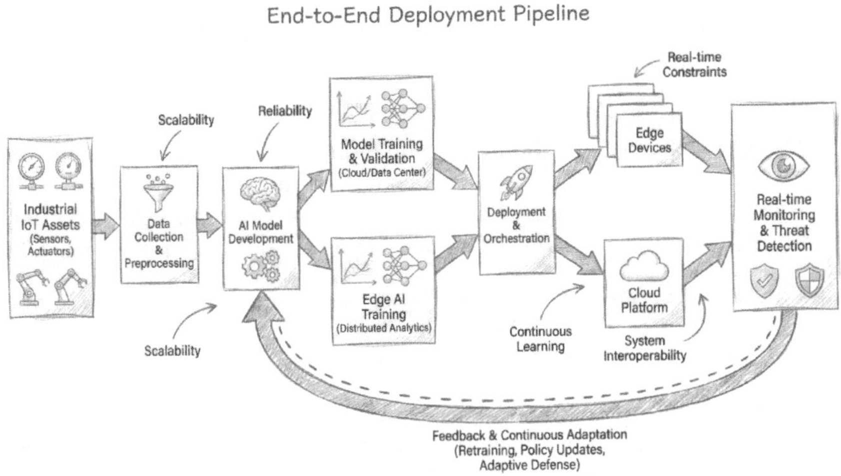

FIGURE 13.4: End-to-End Deployment Pipeline of AI-Driven Cyber-Physical Security Systems.

Bibliography

[1] Batool Alabdullah and Suresh Sankaranarayanan. Optimized ensemble machine learning model for cyberattack classification in industrial iot. *Frontiers in Artificial Intelligence*, 8:1685376, 2025.

[2] Abdullah Alfahaid, Easa Alalwany, Abdulqader M Almars, Fatemah Alharbi, Elsayed Atlam, and Imad Mahgoub. Machine learning-based security solutions for iot networks: A comprehensive survey. *Sensors*, 25(11):3341, 2025.

[3] Abdulmohsen Almalawi. A lightweight intrusion detection system for internet of things: Clustering and monte carlo cross-entropy approach. *Sensors*, 25(7):2235, 2025.

[4] Minodora Badea, Olga Bucovețchi, Adrian V Gheorghe, Mihaela Hnatiuc, and Gabriel Raicu. Maritime industry cybersecurity threats in 2025: Advanced persistent threats (APTs), hacktivism and vulnerabilities. *Logistics*, 9(4):178, 2025.

[5] Changhui Bae, Euteum Choi, and Seongjin Lee. Technologies, applications, and challenges of digital twin across industries: A systematic review of the state-of-the-art literature. *IEEE Access*, 2025.

[6] Vijay Govindarajan, Pratik Patel, Sahil Tripathi, Md Azizul Hoque, and Gautam Siddharth Kashyap. Magic-enhanced keyword prompting for zero-shot audio captioning with clip models. *arXiv preprint arXiv:2509.12591*, 2025.

[7] Izzaldin Izhar, Ali Abdullah, Muhammad Zunnurain Hussain, and Muhammad Zulkifl Hasan. Enhancing iot/iiot intrusion detection: A comparative study of hybrid CNN-LSTM and advanced DNN ML model on edge-iiotset. *Spectrum of Engineering Sciences*, pages 1420–1433, 2025.

[8] Imre Kocsis, Boglárka Burján-Mosoni, and István Balajti. A comprehensive review of key cyber-physical systems, and assessment of their education challenges. *IEEE Access*, 2025.

[9] Jiewu Leng, Kaiwen Zuo, Caiyu Xu, Xueliang Zhou, Shuai Zheng, Jiawen Kang, Qiang Liu, Xin Chen, Weiming Shen, Lihui Wang, et al. Physics-informed machine learning in intelligent manufacturing: a review. *Journal of Intelligent Manufacturing*, pages 1–43, 2025.

[10] Guang Li, Dong Zhao, Yueyang Li, Tieliang Sun, Feng Lv, and Ruoyu Sun. Time series industrial data anomaly detection algorithm based on transformer. In *2025 40th Youth Academic Annual Conference of Chinese Association of Automation (YAC)*, pages 2061–2066. IEEE, 2025.

[11] Fujun Qiu, Ashwini Kumar, Jiang Hu, Poorva Sharma, Yu Bing Tang, Yang Xu Xiang, and Jie Hong. A review on integrating iot, iiot, and industry 4.0: A pathway to smart manufacturing and digital transformation. *IET Information Security*, 2025(1):9275962, 2025.

[12] Rajiv Raich, Anagha Raich, and Nandkishor Kinhekar. Securing the convergence of it and ot networks in cyber physical system: Policy, architecture and implementation challenges. In *ITM Web of Conferences*, volume 79, page 01005. EDP Sciences, 2025.

[13] Anna Reichardt, Matthias Murawski, and Markus Bick. The use of the internet of things to increase energy efficiency in manufacturing industries. *International Journal of Energy Sector Management*, 19(2):369–388, 2025.

[14] Manuel JCS Reis. Ai-driven anomaly detection for securing iot devices in 5g-enabled smart cities. *Electronics*, 14(12):2492, 2025.

[15] Iqbal H Sarker, Md Hasan Furhad, and Raza Nowrozy. Ai-driven cybersecurity: an overview, security intelligence modeling and research directions. *SN Computer Science*, 2(3):173, 2021.

[16] Thandy Simanjuntak. Enhancing cyber-physical system security: A review of detection and mitigation techniques. *International Journal of Information System and Innovative Technology*, 3(2):10–16, 2024.

[17] Arpita Soni, Sahil Tripathi, Gautam Siddharth Kashyap, Manaswi Kulahara, Mohammad Anas Azeez, Zohaib Hasan Siddiqui, Nipun Joshi, and Jiechao Gao. Can we predict your next move without breaking your privacy? *arXiv preprint arXiv:2507.08843*, 2025.

[18] R Tejaswini, P Abinaya, SS Anuprabha, Saranya Karattupalayam Chidambaram, Sudhanshu Arya, Yogesh Kumar Choukiker, and Abhijit Bhowmick. Anomaly detection in iot networks: A deep learning approach using autoencoders. *IEEE Access*, 2025.

[19] Siqi Yang, Wenqiang Pan, Min Li, Mingyong Yin, Hao Ren, Yue Chang, Yidou Liu, Senyao Zhang, and Fang Lou. Industrial internet of things intrusion detection system based on graph neural network. *Symmetry*, 17(7):997, 2025.

[20] Yagmur Yigit, Mohamed Amine Ferrag, Mohamed C Ghanem, Iqbal H Sarker, Leandros A Maglaras, Christos Chrysoulas, Naghmeh Moradpoor, Norbert Tihanyi, and Helge Janicke. Generative ai and llms for critical infrastructure protection: evaluation benchmarks, agentic ai, challenges, and opportunities. *Sensors*, 25(6):1666, 2025.

[21] Tamara Zhukabayeva, Lazzat Zholshiyeva, Nurdaulet Karabayev, Shafiullah Khan, and Noha Alnazzawi. Cybersecurity solutions for industrial internet of things–edge computing integration: Challenges, threats, and future directions. *Sensors*, 25(1):213, 2025.

Chapter 14

Trustworthy LLMs for Mission-Critical Applications: Benchmarks and Gaps

Niharika Jain

14.1 Introduction

The capabilities of Large Language Models (LLMs) in natural language understanding, reasoning, and generation have been impressive and it has thus been applied in mission-critical areas such as healthcare [7], finance [27], defense [5], and emergency management [15]. These environments demand a high level of stakes, ensuring improper, biased, and dangerous model outputs, which could result in dire social, economic, and operational outcomes. Consequently, the problem of the reliability of LLM-based systems has become one of the priorities of research.

Despite impressive performance on general benchmarks LLMs mainly shows limitations related to hallucination [1], lack of transparency [23], susceptibility to adversarial prompts [34] and inconsistent behavior under distribution shifts [25]. These challenges restrict their reliable deployment in safety-sensitive applications. Trustworthy LLMs aim to solve these concerns through robustness [8], interpretability [22], fairness [8] and alignment with human values and regulatory requirements [8]. Figure 14.1 presents an overview of trustworthy LLM frameworks for mission-critical systems by showing the interaction between evaluation, governance, and deployment components.

This chapter surveys existing benchmarks and evaluation methodologies in Section 14.2, discusses trust-enhancement techniques in Section 14.3, analyzes deployment challenges in Section 14.4 and concludes with future research directions in Section 14.5.

DOI: 10.1201/9781003774815-14

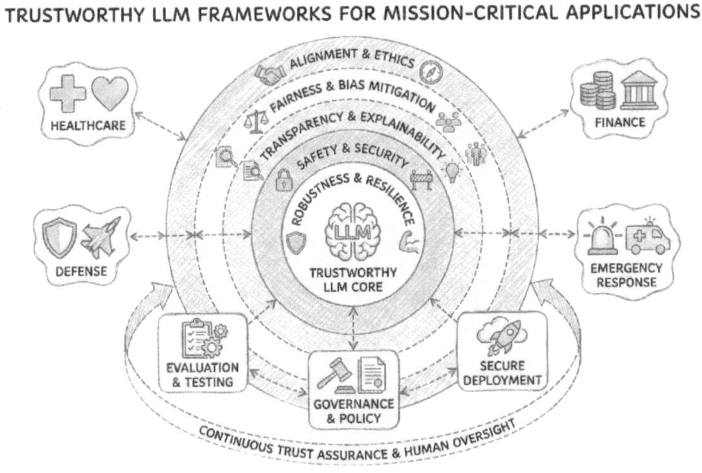

FIGURE 14.1: Overview of Trustworthy LLM Frameworks for Mission-Critical Applications.

14.2 Evaluation Benchmarks and Metrics for Trustworthy LLMs

Systematic evaluation is important for evaluating the reliability and safety of LLMs in mission-critical contexts [7]. Traditional benchmarks like GLUE [29], SuperGLUE [28] and MMLU [12] that mainly focus on linguistic competence and reasoning ability but do not capture any type of safety, performance and ethical dimensions. Recent efforts have introduced specialised benchmarks targeting hallucination detection [13], factual consistency [6], bias assessment [3], and adversarial resilience [24]. There are some examples which include TruthfulQA [19], RealToxicityPrompts [10], AdvGLUE [30] and HELM [18]. These benchmarks evaluate model behavior under adversarial inputs, ambiguous queries and sensitive scenarios.

There are some performance metrics that extend beyond accuracy and perplexity to include calibration error, uncertainty estimation, fairness scores and safety violation rates [26]. Continuous evaluation pipelines are increasingly adopted to monitor model behavior during deployment. Figure 14.2 shows a good benchmarking framework for trustworthy LLM evaluation.

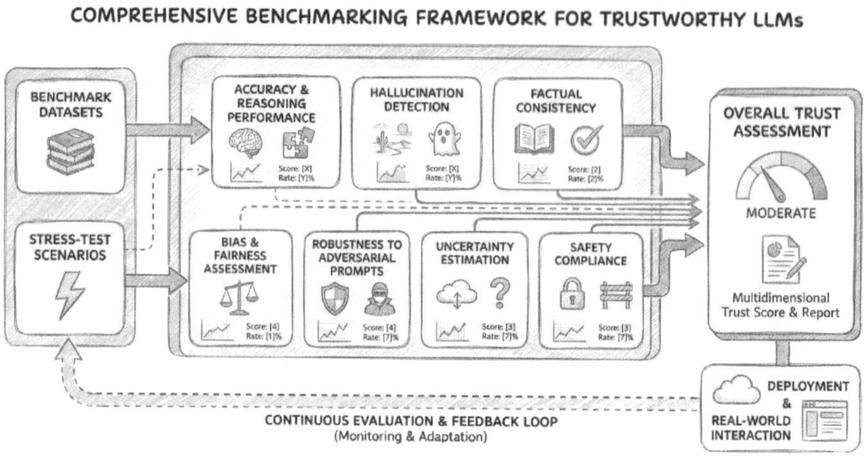

FIGURE 14.2: Comprehensive Benchmarking Framework for Evaluating Trustworthy LLMs.

14.3 Trust Enhancement and Alignment Techniques

Enhancing the trustworthiness of LLMs requires integrating technical, procedural and governance mechanisms throughout the model lifecycle [8]. Alignment techniques aim to secure that model outputs rules to human values [14], domain constraints [20] and regulatory policies [17]. Fine-tuning with human feedback (RLHF) [32], constitutional AI [4] and supervised safety training [31] are widely adopted to reduce harmful behaviours and improve response reliability. Performance is enhanced through adversarial type of training [2], data augmentation [16] and uncertainty-aware modeling [11].

Bias mitigation strategies which includes balanced data curation [9], fairness-aware optimization [33] and post-processing adjustments [21]. Figure 14.3 shows combined architecture for enhancement of the trust and alignment in LLM systems.

14.4 Deployment Challenges, Gaps, and Future Directions

Utilizing reliable LLCM in high risk settings is a challenging technical, organizational, and regulatory problem to implement [7]. The major concerns

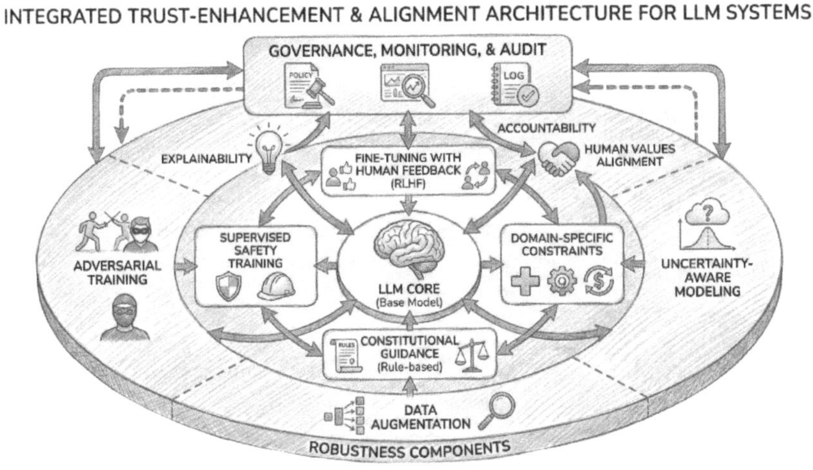

FIGURE 14.3: Integrated Architecture for Trust Enhancement and Alignment in LLM Systems.

are the scalability of the model, privacy of data, explainability, accountability of laws, and the performance limitations in real time. LLMs usually lack the ability to adapt to domains, be long-term consistent, and resilient to rare but critical situations. The certification and regulatory approval are not easily facilitated by limited interpretability and opaque training processes. Moreover, the existing systems of governance are behind the accelerating technological changes.

Hybrid human-AI decision systems are emerging, as well as federated learning to preserve privacy, formal verification, and specialization of models by domain is anticipated to appear in the future of AI and machine learning, as well as federated learning. Figure 14.4 shows the deployment pipeline of the trustworthy LLMs, which involves training and validation, and monitoring and governance.

14.5 Conclusion and Future Works

This chapter has discussed how trustworthy LLMs can be used in mission-critical systems with a focus on evaluation benchmarks, alignment methods, and the issues of deployment. Although current developments have made major steps to enhance the capabilities of the models, there are still considerable gaps in maintaining uniform security, fairness, and reliability.

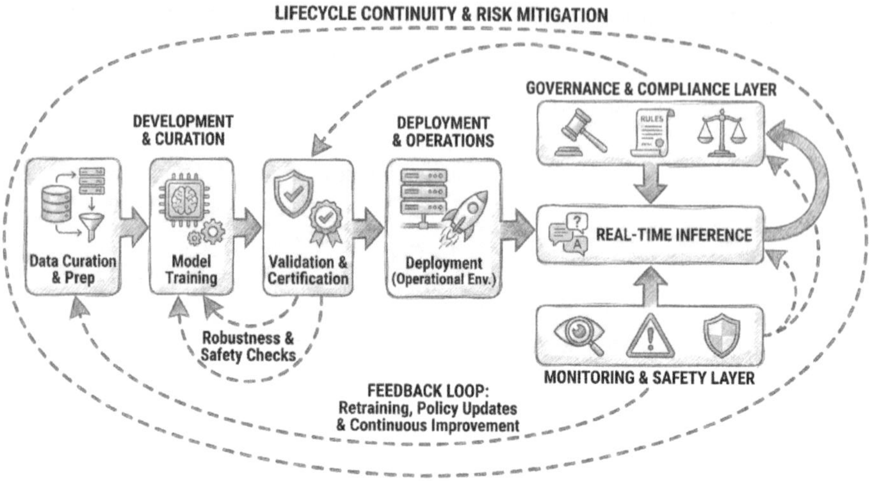

FIGURE 14.4: End-to-End Deployment Pipeline of Trustworthy LLMs for Mission-Critical Systems.

The research should continue in the future with the unified benchmarking standards, interpretable foundation models, effective alignment strategies, and regulator-compliant deployment frameworks. Sustainable adoption will involve the importance of interdisciplinary approach with computer science, ethics, law and domain knowledge.

Bibliography

[1] Aisha Alansari and Hamzah Luqman. Large language models hallucination: A comprehensive survey. *arXiv preprint arXiv:2510.06265*, 2025.

[2] Daniele Angioni, Luca Demetrio, Maura Pintor, Luca Oneto, Davide Anguita, Battista Biggio, and Fabio Roli. Robustness-congruent adversarial training for secure machine learning model updates. *IEEE Transactions on Pattern Analysis and Machine Intelligence*, 2025.

[3] Gemma Galdon Clavell, Rubén González-Sendino, and Paola Vazquez. Demographic benchmarking: Bridging socio-technical gaps in bias detection. *arXiv preprint arXiv:2501.15985*, 2025.

[4] Meade Cleti. Constitutional ai: Principles, practices, and implementation in large language model development. *Framework*, 9:C3AI.

[5] Willer Gomes dos Santos. *USE OF LARGE LANGUAGE MODELS TO SUPPORT AEROSPACE DEFENSE SYSTEMS ENGINEERING.* PhD thesis, Instituto Tecnológico de Aeronáutica, 2025.

[6] Maria Eriksson, Erasmo Purificato, Arman Noroozian, Joao Vinagre, Guillaume Chaslot, Emilia Gomez, and David Fernandez-Llorca. Can we trust ai benchmarks? an interdisciplinary review of current issues in ai evaluation. *arXiv preprint arXiv:2502.06559*, 2025.

[7] Matteo Esposito, Francesco Palagiano, Valentina Lenarduzzi, and Davide Taibi. On large language models in mission-critical it governance: Are we ready yet? In *2025 IEEE/ACM 47th International Conference on Software Engineering: Software Engineering in Practice (ICSE-SEIP)*, pages 504–515. IEEE, 2025.

[8] Md Meftahul Ferdaus, Mahdi Abdelguerfi, Elias Loup, Kendall N. Niles, Ken Pathak, and Steven Sloan. Towards trustworthy ai: A review of ethical and robust large language models. *ACM Computing Surveys*, 58(7):1–43, 2026.

[9] Anna Foka, Gabriele Griffin, Dalia Ortiz Pablo, Paulina Rajkowska, and Sushruth Badri. Tracing the bias loop: Ai, cultural heritage and bias-mitigating in practice. *AI & SOCIETY*, pages 1–13, 2025.

[10] Samuel Gehman, Suchin Gururangan, Maarten Sap, Yejin Choi, and Noah A Smith. Realtoxicityprompts: Evaluating neural toxic degeneration in language models. *arXiv preprint arXiv:2009.11462*, 2020.

[11] Vijay Govindarajan, Pratik Patel, Sahil Tripathi, Md Azizul Hoque, and Gautam Siddharth Kashyap. Magic-enhanced keyword prompting for zero-shot audio captioning with clip models. *arXiv preprint arXiv:2509.12591*, 2025.

[12] Dan Hendrycks, Collin Burns, Steven Basart, Andy Zou, Mantas Mazeika, Dawn Song, and Jacob Steinhardt. Measuring massive multitask language understanding. *arXiv preprint arXiv:2009.03300*, 2020.

[13] Lei Huang, Weijiang Yu, Weitao Ma, Weihong Zhong, Zhangyin Feng, Haotian Wang, Qianglong Chen, Weihua Peng, Xiaocheng Feng, Bing Qin, et al. A survey on hallucination in large language models: Principles, taxonomy, challenges, and open questions. *ACM Transactions on Information Systems*, 43(2):1–55, 2025.

[14] Linus Ta-Lun Huang, Gleb Papyshev, and James K Wong. Democratizing value alignment: From authoritarian to democratic ai ethics. *AI and Ethics*, 5(1):11–18, 2025.

[15] Shumaila Javaid, Hamza Fahim, Bin He, and Nasir Saeed. Large language models for uavs: Current state and pathways to the future. *IEEE Open Journal of Vehicular Technology*, 2024.

[16] Sunwoo Kim, Hun Im, Woojun Lee, Seonggye Lee, and Pilsung Kang. Robustmixgen: Data augmentation for enhancing robustness of visual–language models in the presence of distribution shift. *Neurocomputing*, 619:129167, 2025.

[17] Hariharan Pappil Kothandapani. Ai-driven regulatory compliance: Transforming financial oversight through large language models and automation. *Emerging Science Research*, 12(1):12–24, 2025.

[18] Percy Liang, Rishi Bommasani, Tony Lee, Dimitris Tsipras, Dilara Soylu, Michihiro Yasunaga, Yian Zhang, Deepak Narayanan, Yuhuai Wu, Ananya Kumar, et al. Holistic evaluation of language models. *arXiv preprint arXiv:2211.09110*, 2022.

[19] Stephanie Lin, Jacob Hilton, and Owain Evans. Truthfulqa: Measuring how models mimic human falsehoods. In *Proceedings of the 60th annual meeting of the association for computational linguistics (volume 1: long papers)*, pages 3214–3252, 2022.

[20] Haoran Lu, Luyang Fang, Ruidong Zhang, Xinliang Li, Jiazhang Cai, Huimin Cheng, Lin Tang, Ziyu Liu, Zeliang Sun, Tao Wang, et al. Alignment and safety in large language models: Safety mechanisms, training paradigms, and emerging challenges. *arXiv preprint arXiv:2507.19672*, 2025.

[21] Shaina Mackin, Vincent J Major, Rumi Chunara, and Remle Newton-Dame. Post-processing methods for mitigating algorithmic bias in healthcare classification models: An extended umbrella review. *BMC Digital Health*, 3(1):26, 2025.

[22] Anthony Lawrence Paul. Improving explainability in large language models (LLMs).

[23] Pejman Peykani, Fatemeh Ramezanlou, Cristina Tanasescu, and Sanly Ghanidel. Large language models: A structured taxonomy and review of challenges, limitations, solutions, and future directions. *Applied Sciences*, 15(14):8103, 2025.

[24] Nitin Rane, Suraj Kumar Mallick, and Jayesh Rane. Adversarial machine learning for cybersecurity resilience and network security enhancement. *Available at SSRN 5337152*, 2025.

[25] Alexander Smith, William Martinez, Sophia Garcia, Benjamin Thomas, Olivia Davis, and Weimang Ye. Understanding distribution shift in llms: Methods, evaluations, and challenges.

[26] Arpita Soni, Sahil Tripathi, Gautam Siddharth Kashyap, Manaswi Kulahara, Mohammad Anas Azeez, Zohaib Hasan Siddiqui, Nipun Joshi, and Jiechao Gao. Can we predict your next move without breaking your privacy? *arXiv preprint arXiv:2507.08843*, 2025.

[27] Ahmadreza Tavasoli, Maedeh Sharbaf, and Seyed Mohamad Madani. Responsible innovation: A strategic framework for financial llm integration. *arXiv preprint arXiv:2504.02165*, 2025.

[28] Alex Wang, Yada Pruksachatkun, Nikita Nangia, Amanpreet Singh, Julian Michael, Felix Hill, Omer Levy, and Samuel Bowman. Superglue: A stickier benchmark for general-purpose language understanding systems. *Advances in neural information processing systems*, 32, 2019.

[29] Alex Wang, Amanpreet Singh, Julian Michael, Felix Hill, Omer Levy, and Samuel Bowman. GLUE: A multi-task benchmark and analysis platform for natural language understanding. In Tal Linzen, Grzegorz Chrupała, and Afra Alishahi, editors, *Proceedings of the 2018 EMNLP Workshop BlackboxNLP: Analyzing and Interpreting Neural Networks for NLP*, pages 353–355, Brussels, Belgium, November 2018. Association for Computational Linguistics.

[30] Boxin Wang, Chejian Xu, Shuohang Wang, Zhe Gan, Yu Cheng, Jianfeng Gao, Ahmed Hassan Awadallah, and Bo Li. Adversarial glue: A multi-task benchmark for robustness evaluation of language models. *arXiv preprint arXiv:2111.02840*, 2021.

[31] Wei Wang, Chong Yu, Tengyu Zhang, Feiyu Chen, Yufan Liu, Yaqi Liu, and Zongze Wu. Oversized ore segmentation using sam-enhanced u-net with self-supervised pre-training and semi-supervised self-training. *Expert Systems with Applications*, 285:127980, 2025.

[32] Zhongheng Yang, Aijia Sun, Yushang Zhao, Yinuo Yang, Dannier Li, and Chengrui Zhou. Rlhf fine-tuning of llms for alignment with implicit user feedback in conversational recommenders. In *2025 4th International Conference on Artificial Intelligence, Internet of Things and Cloud Computing Technology (AIoTC)*, pages 587–591. IEEE, 2025.

[33] Guo Yu, Lianbo Ma, Xilu Wang, Wei Du, Wenli Du, and Yaochu Jin. Towards fairness-aware multi-objective optimization. *Complex & Intelligent Systems*, 11(1):50, 2025.

[34] Giulio Zizzo, Giandomenico Cornacchia, Kieran Fraser, Muhammad Zaid Hameed, Ambrish Rawat, Beat Buesser, Mark Purcell, Pin-Yu Chen, Prasanna Sattigeri, and Kush Varshney. Adversarial prompt evaluation: Systematic benchmarking of guardrails against prompt input attacks on llms. *arXiv preprint arXiv:2502.15427*, 2025.

Part IV

Building Reliable and Secure AI Systems

Chapter 15

Bridging Theory and Practice: A LangChain-Based Virtual Assistant for Corporate AI Integration

Vijay Govindarajan

15.1 Introduction

This chapter focuses on the complexities of introducing Artificial Intelligence (AI) projects into the corporate setting, discussing the various obstacles that oftentimes prevent successful adoption of AI in companies where it is trying to be implemented [7, 8, 20]. One of the identified problems is that of the integration gap, as a result of the lack of alignment between theoretical AI developments and their business applications. The reason behind the failure of many AI projects is that of overly optimistic expectations, miscalculated data needs, and lack of scalability. The quality and quantity of data supplied to AI systems are crucial to their functionality, which is why the failure of the project is inevitable. Besides, the AI solutions that are not flexible in various industries also add to the issues of integration further [21]. In an attempt to address this gap, the chapter examines how to come up with low-budget and practical AI proofs of concept, which are cost-effective and operationally versatile. One of the main directions is the development of AI tools that would be available to the employees and would improve the daily working process. An approach here also focuses on the necessity of AI solutions both technically sound and understandable and useful on the operation level. Thus, the chapter is organized with an analysis part, where the theoretical background is presented, a literature review of the available chatbot frameworks-based projects such as LangChain is provided [6], and the methodology and implementation of the project are discussed.

DOI: 10.1201/9781003774815-15

History Chatbots Chatbots have developed into more sophisticated forms, with earlier models relying on straightforward rules, such as ELIZA, an early AI-based chatbot, as discussed by [26] and more recent models, such as GPT [4] which is a Large Language Model (LLM) powered by AI. The revolution of chatbot is due to LLM, which allows more natural and context-driven communication to take place [28]. Their uses are diverse, spreading across education fields, customer service, and medical care areas, where AI-driven chatbots can be used in the learning process, customer support, and even initial medical examination. LangChain technologies also complement the functionality of chatbots, eliminating the necessity to perform fine-tuning regarding the creation of specific models and high computational expenses. Natural Language Processing (NLP) is the intersection of linguistics and computer science that assists computers in comprehending and translating human language and generating it in response to different queries and inputs provided by users through language interfaces [19]. It has since developed into complex deep learning models such as neural networks and transformers over time as opposed to simple systems which rely on rules in decision-making processes [11], customer service [5], and healthcare [24]. The most important aspect of this area is Natural Language Understanding (NLU) [25] which is concerned with the meaning, context, and structure of language. This is particularly relevant in fields such as sentiment analysis [27] and medical research [18], where accurate interpretation matters.

One of the largest problems in AI is to understand human language. NLU enables AI to decode text and speech, yet it cannot handle the nuances of human communication quite well [16, 3] manages the details of human communication, although it fails to recognize it quite well. Language and terminologies can be used in various ways and therefore an AI cannot find the correct meaning without knowing the context. Sentence structure may give rise also to confusion because the same sentence has different meanings upon reading. Moreover, AI should be able to identify the references in the conversation even to pronouns to comprehend the meaning comprehensively. There is an extra difficulty of idioms, sarcasm and humor. Most phrases do not necessarily make the same words in the translated language, and their general meaning relies on the culture [10]. AI must not only analyze words but it should learn more about the way humans communicate. Conversely, Natural Language Generator (NLG) assists artificial intelligence in generating human texts to convert information into meaningful language [13, 9]. With the ongoing development of AI, the advancements in NLU and NLG will result in greater natural and meaningful human-machines interactions.

The chapter will be followed by the next section which will address the pre-implementation phase. The post-implementation results are outlined in Section 15.3. The discussion of the findings is conducted in Section 15.4. The chapter is concluded and followed with future work in Section 15.5.

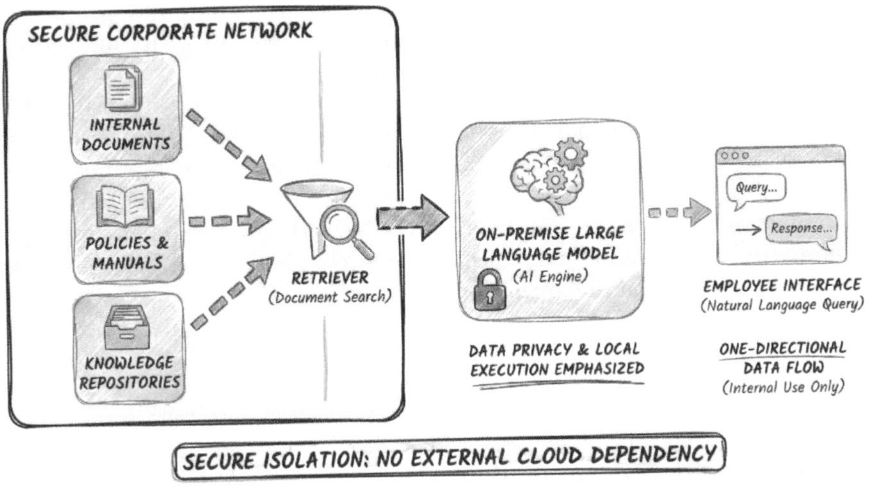

FIGURE 15.1: Pre-Implementation Architecture of the Secure AI-Powered Virtual Assistant for Kratos.

15.2 Pre-Implementations

The study will provide a secure, internet-based AI-driven Virtual Assistant to Kratos and allow the workers to access and comprehend the company information with less effort (see Figure 15.1). The employees do not need to use complicated searches or structured queries to ask questions, but rather use natural language and receive specific and relevant responses. The system consists of two major components the Retriever, which can locate the appropriate documents in a short time, and the Virtual Assistant, which can communicate the content in an easily comprehensible manner, through the advanced AI models, such as Dolly and Mistral. We have an assistant that tracks and executes its work inside the secure network of Kratos, as opposed to cloud-based solutions, and in this way sensitive information remains confidential. It assists in bringing the new employees up to date, aids engineers in technical documentation and makes the administrative workload easy. It is equipped with in-built security, quick processing and is a usability oriented tool that changes how the employees will engage company knowledge. It establishes a precedent in the secure and intelligent access to information by filling the gap between AI and workplace requirements. The future of AI-driven virtual assistants is being made in our project where we are building a local, secure and highly qualified system to suit Kratos. By combining innovative LLCs with other technologies, such as LangChain and Chroma, we are guaranteeing a smooth user-friendly experience without compromising on data privacy and security. As compared to the conventional virtual assistant, ours will be fully within the infrastructure of Kratos, which will present us with full control of the data and its European-compliant nature.

Technically, we are working on optimization of resource-intensive LLMs to execute effectively in local systems with the use of CUDA accelerators and quantization. This guarantees smooth running with computational requirements being within reach. The assistant will be closely connected to the internal systems at Kratos and will make the work of the company more productive due to the simplified information search, the administrative processes, and even help to find documents.

15.3 Post-Implementations

The assistant is being planned to be user friendly and available to all employees irrespective of their technical background (see Figure 15.2). It will make the responses quick and precise and will make the process less frustrating and efficient. The accuracy and relevance of it will be refined constantly with the assistance of a built-in feedback system. The most important obstacles are the optimization of LLMs on local level, the harmonious integration of the technology with the current systems, and making employees interested in the implementation of the technology. We will deal with these by carrying out technical innovations, strict cooperation with the internal teams, and involving training sessions. In conclusion, the chapter is regarding the practicality, security, and value of AI to Kratos and assisting the teams to work smarter and more efficiently. In order to create the Virtual Assistant of Kratos, we have selected the context-provision strategy rather than the fine-tuning of the LLM. Although the process of fine-tuning provides more customization, it has high computational and large-data requirements with frequent updates. On the contrary, the vector-based approach enables more convenient updates, flexibility of models, and efficient configuration with LangChain and Chroma DB. One of the challenges was the integration of internal documentation at Kratos which is represented in different formats. To solve this, we have two approaches, one is the HTML approach to structured documents where hierarchy of headings are used to extract the content and the other is the PDF approach where metadata and table of contents are used to organize the text without losing the context. Both techniques incorporate the use of a close text processing to enhance the system to retrieve the relevant information. To be more precise, we included the use of cosine similarity to compare user queries and the most related documents as well as included additional metadata to enhance retrieval. This can be refined as the project progresses to become an even more sophisticated, responsive and flexible virtual assistant of Kratos

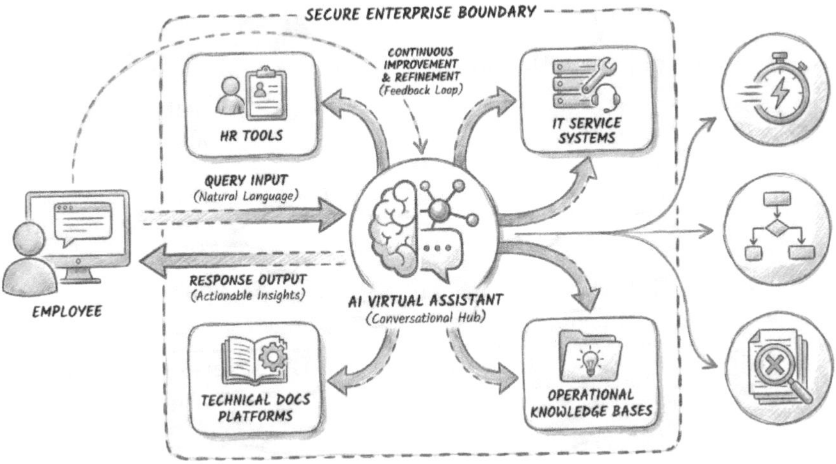

FIGURE 15.2: Post-Implementation Framework Illustrating System Integration and User Interaction Workflows.

[12, 22, 23, 17, 1, 2, 15, 14] and we can envision this as a combination of the approach and fine-tuning as the project advances more and more.

15.4 Findings

The chapter is devoted to streamlining and streamlining the documentation processing and creating a virtual assistant, which corresponds to the needs of the company. The system can read and extract text, tables and figures to the PDFs using the LayoutPDFReader of the LLM Sherpa project and preserve the surrounding of the text which can be embedded back to its proper context. An external model was not taken because privacy, security, and flexibility are required. Sentence Transformers of BERT were used to get fast, high-quality embeddings and Chroma to store data easily. Having tried various models, Falcon and Llama 2 were selected due to their good balance between speed and accuracy (see Tables 15.1 to 15.4). A Retriever app, developed using Streamlit allowed searching documents and visualizing the results. Replacement of Euclidean by cosine similarity enhanced the ability to access information, and users had the opportunity of personalizing results and providing feedback. In the case of the Retriever-Augmented Generation (RAG) model, the emphasis was on the effective operation of LLMs. Other models such as Dolly 2.0 7b

TABLE 15.1: Performance Comparison of Selected LLMs

Model	Response Time (s)	Accuracy Score
Falcon-7B	1.8	0.89
Llama2-7B	2.1	0.91
Mistral-7B	1.6	0.93
Dolly-2.0-7B	2.4	0.86

TABLE 15.2: Memory Optimization and Efficiency of LLM Deployment

Technique	Memory Usage (GB)	Inference Speed
Baseline Model	24.5	0.72
Quantization	14.2	0.85
QLoRA	11.8	0.88
Cached Loading	10.5	0.91

TABLE 15.3: System Reliability and Usability Assessment

Metric	Score (%)	Stability Index
Query Success Rate	94.2	0.92
System Uptime	98.7	0.95
Error Recovery	91.5	0.89
User Satisfaction	92.3	0.93

TABLE 15.4: User Testing and Benchmark Performance Comparison

System	Response Accuracy	User Rating
Kratos Assistant	0.91	4.6/5
ChatGPT	0.93	4.7/5
Internal Search Tool	0.84	4.1/5
Manual Search	0.76	3.8/5

and Mistral 7b have been selected to support the NVIDIA RTX 4090, with methods such as quantization and QLoRA to minimize memory requirements.

The project has provided a robust system with a quick and correct processing of documents and smart virtual assistance that is capable of dealing with complicated questions and providing valuable and context-sensitive answers. To develop an effective virtual assistant with the help of Streamlit and advanced LLPs such as Mistral (see Table 15.1). First, the design of Streamlit, that was designed with the light applications in mind, could not cope with the heavy memory requirements of the LLMs that rapidly exhausted the memory capacity of the GPU. The team resolved this by session state management and the method of st. cache resource. This method also made sure that the model loaded once even when many people would be using the app which meant that efficiency and stability was highly enhanced. Having addressed the memory issues, the emphasis was made on making the app more user-friendly. Chat memory was introduced to ensure that the assistant can recall the previous conversations and users can look through various databases without losing the context. Using such options as the simulation of typing in real-time and the links to the original documents made the interaction more interactive and educational. In addition to the complexity of operating large models in place, methods such as quantization and QLoRA were employed to allow models such as Dolly 2.0 7b and Mistral 7b to be efficiently deployed on the hardware available. These optimizations minimized the memory and guaranteed the smooth operation (see Table 15.2).

The project had its share of challenges, which proved to be an opportunity to improve the system and challenge the capabilities (see Table 15.3). What came out was a stable, intuitive prototype of a virtual assistant that demonstrates how AI could transform the manner in which companies handle and utilize their data. The assistant is able to make sense of complicated queries, search and get the correct information as well as answers in detail with helpful links to the source. Its memory system tracks the past interactions, thus allowing an easy context-sensitive conversation. Although a simple buffer memory is in use at the moment, more sophisticated features will be presented in the future. The process of integration into the Kratos network was not difficult, and several users could use the assistant at the same time. It is very effective in managing diverse collections of documents and matching responses to company rules and rigidities to have correct and compliant information.

The second step is practical testing step of the Kratos Virtual Assistant (see Table 15.4). It is aimed at monitoring its performance in real-life situations, whereby workers in other sections such as engineering and quality management will apply it in their day to day duties. It is planned to divide the users according to their levels of access, starting with those who required access to quality management information and moving to super users who have the entire access. The method of testing will combine both hard data, such as the speed of response and reliability, and the personal information on the extent to which the responses by the assistant are accurate and helpful. The users will be empowered to rate responses promptly and leave comments when something is wrong and this will help to identify areas of improvement. The performance of an assistant, as well, is compared with such tools as ChatGPT to ensure the performance is industry-standard without sacrificing the needs peculiar to Kratos. This stage will examine the behavior of the assistant with multiple users, scalability, as well as under heavy use.

15.5 Conclusions

In this chapter, the LangChain framework is used to create a virtual assistant, which uses state-of-the-art NLP and transformer models such as Mistral 7b. The assistant operates on the premise of converting documents into embeddings and providing pertinent responses on the basis of cosine similarity searches. Optimization of performance on local hardware was one of the primary issues, especially when using the Mistral model in an NVIDIA RTX 4090. The other hassle was the development of an individual virtual environment. Irrespective of these challenges, the assistant is economical and consumes few resources with scaleable and flexible solutions. Privacy of the data is also of primary concern and the assistant is hosted locally and runs on open-source software. This chapter demonstrates that it is possible to introduce LLMs to business settings without breaking the bank, and companies can proceed to improve their productivity with the use of AI. It also provides new opportunities of AI usage in the fields where technology is not a major focus, and it can be automated in the future.

Bibliography

[1] Gopich Bandarupalli. Advancing smart transportation via ai for sustainable traffic solutions in saudi arabia's. In *2025 First International Conference on Advances in Computer Science, Electrical, Electronics, and Communication Technologies (CE2CT)*, pages 470–475. IEEE, 2025.

[2] GOPICHAND BANDARUPALLI. Enhancing microservices performance with ai-based load balancing: A deep learning perspective. 2025.

[3] Emily M Bender and Alexander Koller. Climbing towards nlu: On meaning, form, and understanding in the age of data. In *Proceedings of the 58th annual meeting of the association for computational linguistics*, pages 5185–5198, 2020.

[4] Tom Brown, Benjamin Mann, Nick Ryder, Melanie Subbiah, Jared D Kaplan, Prafulla Dhariwal, Arvind Neelakantan, Pranav Shyam, Girish Sastry, Amanda Askell, et al. Language models are few-shot learners. *Advances in neural information processing systems*, 33:1877–1901, 2020.

[5] Erik Brynjolfsson, Danielle Li, and Lindsey Raymond. Generative ai at work. *The Quarterly Journal of Economics*, 140(2):889–942, 2025.

[6] Harrison Chase. Langchain. `https://github.com/langchain-ai/langchain`, 2022.

[7] Francois Chollet and François Chollet. *Deep learning with Python*. simon and schuster, 2021.

[8] Yogesh K Dwivedi, Laurie Hughes, Elvira Ismagilova, Gert Aarts, Crispin Coombs, Tom Crick, Yanqing Duan, Rohita Dwivedi, John Edwards, Aled Eirug, et al. Artificial intelligence (ai): Multidisciplinary perspectives on emerging challenges, opportunities, and agenda for research, practice and policy. *International journal of information management*, 57:101994, 2021.

[9] Albert Gatt and Emiel Krahmer. Survey of the state of the art in natural language generation: Core tasks, applications and evaluation. *Journal of Artificial Intelligence Research*, 61:65–170, 2018.

[10] Daniel Hershcovich, Stella Frank, Heather Lent, Miryam De Lhoneux, Mostafa Abdou, Stephanie Brandl, Emanuele Bugliarello, Laura Cabello Piqueras, Ilias Chalkidis, Ruixiang Cui, et al. Challenges and strategies in cross-cultural nlp. *arXiv preprint arXiv:2203.10020*, 2022.

[11] Enkelejda Kasneci, Kathrin Seßler, Stefan Küchemann, Maria Bannert, Daryna Dementieva, Frank Fischer, Urs Gasser, Georg Groh, Stephan Günnemann, Eyke Hüllermeier, et al. Chatgpt for good? on opportunities and challenges of large language models for education. *Learning and individual differences*, 103:102274, 2023.

[12] Patrick Lewis, Ethan Perez, Aleksandra Piktus, Fabio Petroni, Vladimir Karpukhin, Naman Goyal, Heinrich Küttler, Mike Lewis, Wen-tau Yih, Tim Rocktäschel, et al. Retrieval-augmented generation for knowledge-intensive nlp tasks. *Advances in neural information processing systems*, 33:9459–9474, 2020.

[13] Junyi Li, Tianyi Tang, Wayne Xin Zhao, Jian-Yun Nie, and Ji-Rong Wen. Pre-trained language models for text generation: A survey. *ACM Computing Surveys*, 56(9):1–39, 2024.

[14] Satish Mandavalli. Enhancing dengue outbreak predictions using machine learning: A comparative analysis of models. 2024.

[15] Satish Mandavalli. Enhancing precision: Unveiling individualized treatment effects with advanced computational methods. 2024.

[16] Melanie Mitchell and David C Krakauer. The debate over understanding in ai's large language models. *Proceedings of the National Academy of Sciences*, 120(13):e2215907120, 2023.

[17] Lakshmi Narasimha Raju Mudunuri, Praveen Kumar Maroju, and Venu Madhav Aragani. Leveraging nlp-driven sentiment analysis for enhancing decision-making in supply chain management. In *2025 Fifth International Conference on Advances in Electrical, Computing, Communication and Sustainable Technologies (ICAECT)*, pages 1–6. IEEE, 2025.

[18] Harsha Nori, Nicholas King, Scott Mayer McKinney, Dean Carignan, and Eric Horvitz. Capabilities of gpt-4 on medical challenge problems, 2023.

[19] Daniel W Otter, Julian R Medina, and Jugal K Kalita. A survey of the usages of deep learning for natural language processing. *IEEE transactions on neural networks and learning systems*, 32(2):604–624, 2020.

[20] Andrei Paleyes, Raoul-Gabriel Urma, and Neil D Lawrence. Challenges in deploying machine learning: a survey of case studies. *ACM computing surveys*, 55(6):1–29, 2022.

[21] Nithya Sambasivan, Shivani Kapania, Hannah Highfill, Diana Akrong, Praveen Paritosh, and Lora M Aroyo. "everyone wants to do the model work, not the data work": Data cascades in high-stakes ai. In *proceedings of the 2021 CHI Conference on Human Factors in Computing Systems*, pages 1–15, 2021.

[22] Parth Saxena. Evaluating machine learning algorithms for effective credit card fraud detection. In *2025 10th International Conference on Signal Processing and Communication (ICSC)*, pages 375–380. IEEE, 2025.

[23] Parth Saxena and Janak Devashish. Sentiment analysis and topic modeling of patient reviews on psychiatric medications using cloud-based architecture. In *2025 International Conference on Computational, Communication and Information Technology (ICCCIT)*, pages 892–897. IEEE, 2025.

[24] Arun James Thirunavukarasu, Darren Shu Jeng Ting, Kabilan Elangovan, Laura Gutierrez, Ting Fang Tan, and Daniel Shu Wei Ting. Large language models in medicine. *Nature medicine*, 29(8):1930–1940, 2023.

[25] Alex Wang, Amanpreet Singh, Julian Michael, Felix Hill, Omer Levy, and Samuel Bowman. Glue: A multi-task benchmark and analysis platform for natural language understanding. In *Proceedings of the 2018 EMNLP workshop BlackboxNLP: Analyzing and interpreting neural networks for NLP*, pages 353–355, 2018.

[26] Joseph Weizenbaum. Eliza—a computer program for the study of natural language communication between man and machine. *Communications of the ACM*, 9(1):36–45, 1966.

[27] Lei Zhang, Shuai Wang, and Bing Liu. Deep learning for sentiment analysis: A survey. *Wiley interdisciplinary reviews: data mining and knowledge discovery*, 8(4):e1253, 2018.

[28] Wayne Xin Zhao, Kun Zhou, Junyi Li, Tianyi Tang, Xiaolei Wang, Yupeng Hou, Yingqian Min, Beichen Zhang, Junjie Zhang, Zican Dong, et al. A survey of large language models. *arXiv preprint arXiv:2303.18223*, 1(2), 2023.

Chapter 16

Uncovering Non-Linear Drivers of Equity Valuation Across Sectors

Srinivas Raju Gottimukkala

16.1 Introduction

One of the most significant and hardest task in the current financial markets in a rapid movement is the valuation of a company. The valuation of a company is not only an indication of its current financial position but also defines how investors spend their money and how portfolios are traded, as well as the way a merger or acquisition is negotiated [12]. Valuation in a sense is a map, one that is used to make decisions that can have an industry-wide effect. Just as it has been stated by the authors of the paper by[4] such a concept as value is what defines measurement in a market economy, which explains how central it is in the context of the world of finance. In this chapter, the authors built on the findings of their predecessors such as [5] who had conducted a study on what determines the valuation of companies in various industries. Though their results were useful, much of the literature based on such models tended to rely on linear regression models [15] which, although convenient, may occasionally fail to identify more complicated, real world dynamics determining financial behavior. Thus, this chapter tries to take the needle one step further, employing more up-to-date instruments, namely a Machine Learning (ML) algorithm known as Gradient-Boosted Decision Trees (GBDT). As compared to conventional linear models, GBDTs are more efficient in identifying non-linear relationships and the subtle connections among variables [10].

This chapter dwells upon a broad group of variables, namely on commodity prices, macroeconomic, and company fundamentals to gain insight into what actually causes equity valuations in various industries, according to the Global Industry Classification Standard (GICS). Linear regression and GBDT models were both analyzed and compared through the analysis of monthly stock returns (in this case, used as a proxy of changing the valuation) by means of linear regression and the GBDT model analysis [13] [7]. These findings were

DOI: 10.1201/9781003774815-16

evident that the ML models performed better than the linear ones in that they exhibited a higher capacity to explain returns and valuations variations. This is a call to the increased flexibility and adaptability of the modern financial analysis methods towards a more responsive modeling approach. Among the most valuable findings of the research, one can spot the fact that the significance of various valuation drivers is rather diverse across industries. That is to say that what is important to a tech company may have absolutely nothing to do with what creates value in the energy, health, or financial services industry. An example of this is that the price of commodities may be high in industries such as energy or materials and company-specific fundamentals and economic signals more in financials or consumer discretionary. The awareness of these differences will enable investors and analysts to be more conscious about the signals they should pay more attention on in specific situations, and therefore, make smarter and more specific decisions. The implications of these insights on commonly applied valuation methods such as Discounted Cash Flow (DCF) analysis or a similar company analysis are also practical in nature. The techniques are quite dependent on assumptions [3] [2] [8] regarding growth, any discount rates and any comparisons in the market and the more informed the assumptions are, the more accurate the valuation will be. This chapter can be used to narrow down those assumptions by determining the variables that are the most important to individual sectors and thus make the valuation models more specific and accurate. The next important lesson of this chapter is that it should not be evaluated solely according to previous data or those of the standard financial ratios.

The chapter is as follows the pre-implementation phase will be covered in the next section. Section 16.3 has discussed what happens after the implementation. In Section 16.4 the experimental analysis has carried out. Section 16.5 concludes the chapter with some conclusions and future work.

16.2 Pre-Implementations

The knowledge of what makes a company worth valuing is very fundamental to any person in the financial markets be it the investor, analyst or a corporate strategist (see Figure 16.1). In this section, one inspects the round neck of the determinants of stock valuations in different industries through ten years of data and integrating conventional financial technologies with the potent ML methods. In constructing a reliable data set, this chapter collected historical stock prices on a daily basis, quarterly financial reports (10-Q filings), and earnings call schedules of companies included in the S&P 500, 400, and 600 indexes. Earnings call information was obtained via Stocktwits API and stock price and financial data were obtained via Yahoo Finance. Further information on the prices of commodities and economic variables were retrieved in the

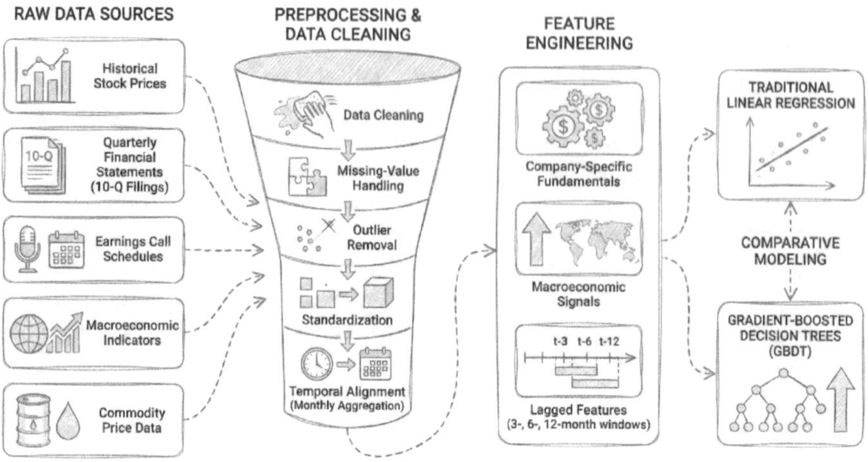

FIGURE 16.1: Pre-Implementation Data Collection and Modeling Framework for Stock Valuation Analysis.

Yahoo Finance and the Federal Reserve Economic Data (FRED) database, which is used to create a more balanced impression of what can affect the valuation [14]. The monthly percentage change in the stock price is the center of the chapter and it is used as a proxy of the change in the value of a company over a given period. It is a straight forward concept that as long we can comprehend the factors that drive stock prices up and down, we can be closer to a deeper insight of what drives valuation. The variables that are taken into account (the independent variables) may include company-specific (such as revenue and profitability) or more general (such as commodity prices and changes in the economy on the whole).

The study spans an analysis of 10 years of time between January 2014 and February 2024, which gives a detailed timeline to the short-term market changes and the long-term valuation trends. Because most of the company and economic indicators are reported on a quarterly basis, monthly percent changes had been calculated in the chapter relative to the most recent releases. The raw data was pre-modeled by cleaning up. This involved finding and dropping outliers such as nonunion stock returns due to one-time events such as the GameStop short squeeze or correction of missing and inconsistent data. In order to maintain consistency, the values representing missing values (in case of absence of earnings calls or data releases) were entered as zeros which means no change has occurred. Variables were then put under standardization in order to compare them in a better way. More was done to manage the natural seasonality in financial reporting to incorporate lagged features (3, 6 or 12 months) using sector-specific patterns. The independent variables were identified using a clustering algorithm called DBSCAN to flag outliers and eliminate them using the algorithmic method of DBSCAN [1]. Two modeling methods were applied to reveal the drivers of stock valuation a simple multivariate linear regression model, and a more sophisticated GBDT model (LightGBM) [6] [11] [9]. They both received sector specific data training to achieve the specific dynamics at play in various industries. To make sure that the results were reliable, the chapter employed 5-fold cross-validation that divides the data into subsets to compute the accuracy of the models on hidden data. These subsets were also designed with the consideration that there was a reasonable proportion of earnings release months in each of these subsets when the stock prices are highly sensitive to new information. Backward stepwise regression process has been used to refine the models and eliminate redundancy between variables. This was through the selection of groups of highly correlated variables and only a single variable of each group retained that gave the best predictive power. All the variables that failed to enhance a better model were dropped out completely after this filtering. Lastly, the two models were streamlined by using cross-validated hyperparameter tuning so that they found the optimal balance between accuracy and simplicity.

16.3 Post-Implementations

This section pays a closer attention to the actual factors of fluctuations in stock values in various sectors (see Figure 16.2). The analysis identifies both simple and complex relationships in the data through the utilization of both a linear regression and a LightGBM. The SHAP values were used to understand the models- the values break down the extent to which each variable affects the stock price changes. By concentrating on models that perform in the middle (according to cross-validation) the overfitting angle can be eliminated, and a more balanced point of view will be presented. The models are constructed with monthly data because they are built monthly, therefore variables that are updated often such as commodity prices or economic indicators would be more influential. Conversely, the company basics (such as financial reports), which are published on a quarterly basis, are not always rated as highly significant, although they are important in the long-term perspective. In interpreting the results therefore it is important to note the frequency of each variable which is being reported. Naturally, the VIX (the market volatility indicator) and the federal funds rate were always in the list of the leading drivers in almost all sectors. These indicators represent the mood of the whole market and financial state of affairs, which by definition influences the way investors price companies. Other influences that appeared quite often included inflation, housing activity, and interest rate spreads.

Considering particular industries: the oil and gas prices dominated in the energy sector, which is no wonder considering the close dependencies of the

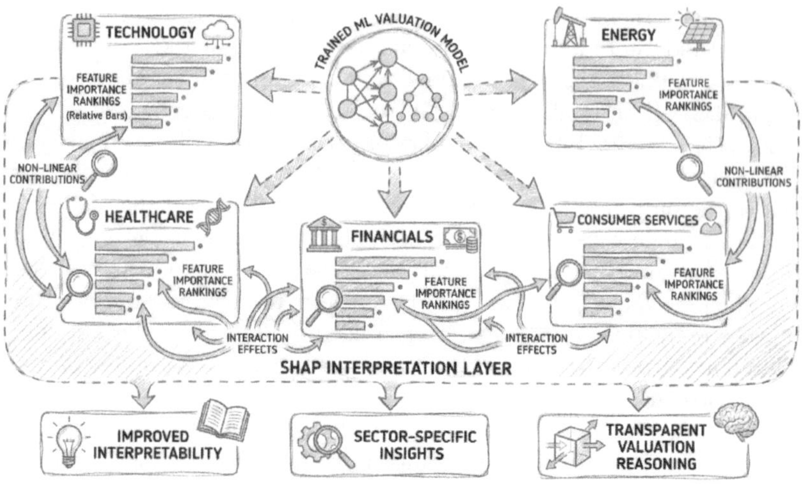

FIGURE 16.2: Post-Implementation Framework Illustrating Sector-Wise Valuation Drivers and Model Interpretation Using SHAP.

industry to the respective markets. Commodity prices and debt related ratios were particularly applicable in materials. Raw materials such as copper influenced the industrials and the housing market trends did the same. In the case of consumer discretionary, commodity inputs, interest rates, and housing starts, it was large indicating the sensitivity of the sector to economic cycles and consumer spending. Concisely, the stock prices are influenced by a combination of both the economic indicators and historical performance of the company- and the mix is quite different in different industries. Although every industry responds differently to the market, there are certain recurring trends when it comes to financial well being, macroeconomic trends and shareholders attitudes.

Lower inflation is significant in consumer goods and services sector which includes furniture, appliances, and home improvement products. Consumers are likely to be more assured of their expenditures and more costs on non-essential items when prices are stable. This is also increased by the government spending which increases the activity in the economy. Financial strength is indicated by high profit margins, minimal debt, increasing amount of free cash flow, and good returns on equity which is quite appreciated on the company level. The consumer staples industry which comprises the necessities such as food, drinks, and household goods demonstrated less model prediction. This is probably because of the variety of products it encompasses. Nevertheless, the main factors that do this are stable commodity prices, reduced interest rates, and decreased market volatility, all of which have kept the cost of production within reach and it is promoting consumer expenditure. Financially, an increase in margins and cash flow are key success indicators.

16.4 Experimental Analysis

This part examines the motivation of stock performance in various areas of the economy- and to what extent ML models can be used to make us know these dynamics (see Tables 16.1 to 16.3). By contrasting a classic linear regression model with the more sophisticated LightGBM the study demonstrates the extent to which ML may be invaluable in understanding the market intricacies. The LightGBM model performed better than the linear model across the board providing a better understanding of what affects stock returns in each sector as presented in Tables 16.1 to 16.3. It increased the adjusted R2 (a measure of the extent of variation of returns explained by the model) more than 12 percent on average. That is a very big leap mostly because LightGBM is able to process the non-linear and usually byzantine affiliations between financial metrics and market conduct, which the linear models cannot manage. The LightGBM model provided more accurate predictions in addition to explaining returns better. It consistently recorded less errors, in terms of Mean Squared error (MSE) and Mean Absolute error (MAE), which is, it was more effective in its predictions of stock movements. It also fared better statistically with higher results in 9 of the 11 sectors that it was assessed in based on F-statistic. Although the two models were statistically sound in general (with very low p-values), LightGBM was more reliable. Regardless, there were some challenges encountered by both models. Their residuals exhibited heteroscedasticity (variable variability) and non-normality (not bell-shaped) which may influence the precision. They aren't deal killers, but they do warn us that there is no model that is flawless, particularly in the world of the sloppy, real-world finance. Low inflation and government spending stimulated consumer confidence and consumer spending in consumer discretionary (think furniture, appliances and home improvement). Investors gave ratings to companies that had high margins, low debts, and good free cash flow. Consumer staples industry (including food and cleaning products) was doing very well in response to the low interest rates and the lack of fluctuations in commodity prices. It is more difficult to model because it encompasses a number of goods; however, such fundamentals as profit margins and cash flow growth

TABLE 16.1: Performance of Linear Regression Model Across Economic Sectors

Sector	Mean Absolute Error	Adjusted R^2
Consumer Discretionary	0.082	0.41
Consumer Staples	0.075	0.45
Health Care	0.091	0.38
Financial Services	0.078	0.44
Information Technology	0.085	0.42

TABLE 16.2: Performance of LightGBM Model Across Economic Sectors

Sector	Mean Absolute Error	Adjusted R^2
Consumer Discretionary	0.061	0.56
Consumer Staples	0.054	0.58
Health Care	0.067	0.51
Financial Services	0.059	0.57
Information Technology	0.063	0.55

TABLE 16.3: Comparative Improvement of LightGBM over Linear Regression

Sector	MAE Reduction (%)	R^2 Improvement
Consumer Discretionary	25.6	+0.15
Consumer Staples	28.0	+0.13
Health Care	26.4	+0.13
Financial Services	24.4	+0.13
Information Technology	25.9	+0.13

were important. The complexity of the health care sector made it more difficult to predict because of the sensitivity of the regulation. Nevertheless, the model discovered that the high profit margins, increase in revenues and good credit conditions favored the stock performance. Economic data such as housing starts and price of commodities were the most significant in the financial services industry except banks and insurance companies. Companies that had a good cash flow, high profit margins and invested in growth were taken positively. Information technology industry was more responsive to innovation and internal effectiveness rather than general market fluctuation.

16.5 Conclusion and Future Works

This chapter gets a little more specific on what is actually behind equity valuations in various sectors and it turns out that it is not a one-shoe-fits-all answer. The analysis did reveal the patterns that the traditional linear models failed to show by using a more flexible ML model of GBDT. It is important to note that in 9 out of 11 industries, the free cash flow was a significant indicator, with profitability ratios coming in second place, which demonstrates the significance of the capacity of the company to raise cash and ensure good financial results. Interestingly, such metrics as net income and EPS, which were believed to be the core, contributed much less. Such insights can assist the investors to perfect their valuation models, make more refined investment decisions, and risk management. To the future, it may be more accurate to introduce new information, such as analyst predictions, corporate directions, or even market perceptions through news and social media. One can also take the great chance to learn more about particular industries or even local markets to get to know the region-specific trends and drivers.

Bibliography

[1] Shadi Abudalfa. *Metaheuristic Clustering Algorithm.* PhD thesis, The Islamic University of Gaza, Gaza, Palestine, 2010.

[2] Aleksanteri Aho. Practical challenges in the application of the discounted cash flow (DCF) model. 2025.

[3] Kai Cao and Haifeng You. Fundamental analysis via machine learning. *Financial Analysts Journal*, 80(2):74–98, 2024.

[4] Jon Danielsson, Kevin R James, Marcela Valenzuela, and Ilknur Zer. Model risk of risk models. *Journal of Financial Stability*, 23:79–91, 2016.

[5] Victor DeMiguel, Lorenzo Garlappi, and Raman Uppal. Optimal versus naive diversification: How inefficient is the 1/n portfolio strategy? *The review of Financial studies*, 22(5):1915–1953, 2009.

[6] Jie Ding and Nigel Meade. Forecasting accuracy of stochastic volatility, garch and ewma models under different volatility scenarios. *Applied Financial Economics*, 20(10):771–783, 2010.

[7] Dustin G Gibson, Tigest Tamrat, and Garrett Mehl. The state of digital interventions for demand generation in low-and middle-income countries: considerations, emerging approaches, and research gaps. *Global Health: Science and Practice*, 6(Supplement 1):S49–S60, 2018.

[8] Vijay Govindarajan, Pratik Patel, Sahil Tripathi, Md Azizul Hoque, and Gautam Siddharth Kashyap. Magic-enhanced keyword prompting for zero-shot audio captioning with clip models. *arXiv preprint arXiv:2509.12591*, 2025.

[9] Yuxuan Huang, Luiz Fernando Capretz, and Danny Ho. Machine learning for stock prediction based on fundamental analysis. In *2021 IEEE Symposium Series on Computational Intelligence (SSCI)*, pages 01–10. IEEE, 2021.

[10] Arif Hussain, Amir Hosein Sakhaei, and Mahmood Shafiee. Machine learning-based constitutive modelling for material non-linearity: A review. *Mechanics of Advanced Materials and Structures*, pages 1–19, 2024.

[11] Guolin Ke, Qi Meng, Thomas Finley, Taifeng Wang, Wei Chen, Weidong Ma, Qiwei Ye, and Tie-Yan Liu. Lightgbm: A highly efficient gradient boosting decision tree. *Advances in neural information processing systems*, 30, 2017.

[12] Khwazi Magubane and Precious Mncayi-Makhanya. The dynamic influence of the global financial cycle: synchronization and impact on south africa's financial cycle. *Cogent Economics & Finance*, 13(1):2451051, 2025.

[13] FR Oswald AJ Mascarenhas, SJ. Artificial intelligence and the emergent turbulent markets: New challenges to corporate ethics today. In *Corporate Ethics for Turbulent Markets: The Market Context of Executive Decisions*, pages 215–242. Emerald Publishing Limited, 2018.

[14] Meri Papavangjeli and Adam Geršl. Monetary policy, macro-financial vulnerabilities, and macroeconomic outcomes. Technical report, IES Working Paper, 2024.

[15] Arpita Soni, Sahil Tripathi, Gautam Siddharth Kashyap, Manaswi Kulahara, Mohammad Anas Azeez, Zohaib Hasan Siddiqui, Nipun Joshi, and Jiechao Gao. Can we predict your next move without breaking your privacy? *arXiv preprint arXiv:2507.08843*, 2025.

Chapter 17

AI-Driven Risk Modeling and Stress Testing in Financial Services

Mohammad Anas Azeez

17.1 Introduction

There are some financial institutions that operate in highly uncertain environments characterized by market volatility [14], credit risk [9], liquidity fluctuations [4] and systemic shocks [11]. There are some good risk modeling and stress testing are therefore important for maintaining financial stability and regulatory compliance. Traditional risk evaluation methods trusts on statistical assumptions [6] and historical type of patterns [15] which mainly fail to capture complex nonlinear dependencies and extreme market events.

There are some recent advances in Artificial Intelligence (AI) and machine learning that have transformed risk analytics by enabling adaptive learning, high-dimensional data processing and scenario-based forecasting [5]. AI-driven frameworks support real-time monitoring, early warning systems [19] and dynamic stress testing under adverse economic conditions [2]. This chapter mainly explores modern AI-based risk modeling approaches and their application in financial services. Table 17.1 shows a comparative overview of traditional and AI-driven risk assessment methods used in banking and investment institutions.

This chapter surveys AI-based risk modeling techniques in Section 17.2, examines stress testing frameworks and scenario analysis in Section 17.3, discusses deployment challenges and future research directions in Section 17.4 and concludes with key data in Section 17.5.

DOI: 10.1201/9781003774815-17

TABLE 17.1: Traditional vs. AI-Based Risk Assessment Methods

Method	MainStrength	Limitation
StatisticalModels	Interpretability	Nonlinearity
EconometricModels	PolicyAnalysis	Rigidity
NeuralNetworks	PatternLearning	Opacity
GradientBoosting	Accuracy	Sensitivity
EnsembleModels	Robustness	Complexity
BayesianModels	Uncertainty	Computation
SupportVector	Generalization	Scalability
KNearestNeighbor	Simplicity	Inefficiency
DecisionTrees	Transparency	Overfitting
HiddenMarkov	Sequencing	Assumptions
Autoencoders	Compression	Interpretability
GaussianProcess	Probabilistic	Scalability
FuzzySystems	Reasoning	Subjectivity
ReinforcementLearning	Adaptation	Instability
NaiveBayes	Speed	Independence
LogisticRegression	Classification	Linearity
PrincipalComponent	Reduction	InformationLoss
PartialLeastSquares	Correlation	Simplification
ExtremeLearning	Speed	Stability
GraphModels	RelationalLearning	Complexity
RuleBasedSystems	Explainability	Rigidity
MarkovChains	Prediction	Memoryless
SparseModels	Parsimony	Underfitting
KernelMethods	Flexibility	Computation
MetaLearning	Adaptability	Overhead

17.2 AI-Based Risk Modeling Techniques

AI-driven risk modeling uses machine learning [5], deep learning [17] and ensemble techniques to capture nonlinear type of patterns in financial data [19]. There are some common approaches which include gradient boosting models [21], neural networks [2], Bayesian inference [3] and reinforcement learning [18].

These models uses various data sources which includes transaction records, macroeconomic indicators, market sentiment and balance sheet information. Feature engineering and dimensionality reduction techniques further increase predictive performance [1]. Compared to linear regression and econometric models [13] AI approaches shows superior and good accuracy and adaptability. Table 17.2 shows the predictive performance of selected AI models for financial risk estimation.

TABLE 17.2: AI Models for Financial Risk Estimation

Model	ApplicationArea	KeyAdvantage
LinearRegression	BaselineModeling	Explainability
RandomForest	CreditRisk	Robustness
NeuralNetwork	MarketForecasting	Nonlinearity
LightGBM	DefaultPrediction	Efficiency
HybridEnsemble	SystemicRisk	Stability
SupportVectorMachine	VolatilityModeling	Precision
BayesianNetwork	DependencyAnalysis	Probabilistic
Autoencoder	FraudDetection	Compression
ConvolutionalNetwork	PatternRecognition	FeatureLearning
RecurrentNetwork	TimeSeries	Memory
Transformer	SequenceModeling	Attention
GaussianProcess	RiskForecasting	Uncertainty
ClusteringModel	Segmentation	Discovery
IsolationForest	AnomalyDetection	Sensitivity
DeepBeliefNetwork	RepresentationLearning	Hierarchy
ExtremeGradientBoosting	CreditScoring	Speed
CapsuleNetwork	PatternExtraction	Robustness
SelfOrganizingMap	Visualization	Topology
SparseAutoencoder	FeatureSelection	Compactness
TemporalConvolution	SequenceAnalysis	Parallelism
GraphNeuralNetwork	NetworkRisk	Dependency
MetaLearner	RapidAdaptation	Transferability
ProbabilisticProgramming	ScenarioAnalysis	Interpretability
MultiTaskLearning	PortfolioModeling	Generalization
FederatedLearning	PrivacyPreservation	Decentralization

17.3 Stress Testing Frameworks and Scenario Analysis

Stress testing is a test of the strengths of financial institutions in bad economic times like recession, liquidity crises and market crash [7]. Stress testing systems that are built on AI create lifelike situations by simulating economic variables and financial portfolios interactions in a realistic manner [16]. Stress scenarios are generally generated in a deep generative model, although Monte Carlo simulations can also be applied, and ensemble models can be used to estimate the sensitivity of a portfolio. These methods make possible the continuous stress testing and adaptive risk management. Reverse stress testing with the assistance of the AI-driven systems is also determined, which can result in institutional failure. Table 17.3 shows the results of the stress testing in various unfavorable conditions.

TABLE 17.3: Common Stress Testing Scenarios

Scenario	RiskSource	Purpose
MildRecession	EconomicSlowdown	Warning
SevereRecession	GDPDecline	Resilience
LiquidityCrisis	FundingShortage	Planning
MarketCrash	AssetCollapse	Assessment
SystemicShock	SectorFailure	Stability
InflationSpike	PriceSurge	Hedging
InterestShock	RateHikes	Protection
CurrencyCrisis	Devaluation	Exposure
EnergyShortage	SupplyDisruption	Continuity
PandemicOutbreak	HealthEmergency	Preparedness
CyberAttack	NetworkBreach	Security
GeopoliticalConflict	Sanctions	RiskControl
HousingCollapse	MortgageDefaults	Mitigation
CommodityCrash	PriceCollapse	Diversification
ClimateDisaster	EnvironmentalRisk	Recovery
BankingCrisis	Insolvency	Safeguarding
TradeWar	Tariffs	Adjustment
TechnologyFailure	SystemOutage	Redundancy
SupplyChainBreakdown	LogisticsDisruption	Resilience
SovereignDefault	DebtCrisis	Containment
LaborStrike	WorkforceDisruption	Continuity
CyberEspionage	DataTheft	Prevention
RegulatoryShock	PolicyChange	Compliance
SpeculationBubble	Overvaluation	Correction
WaterScarcity	ResourceDepletion	Sustainability

17.4 Deployment Challenges and Future Directions

Regardless of the benefits, the implementation of AI-based risk systems is accompanied by the issues of data quality [20], model interpretability [12], regulatory acceptance [8] and cybersecurity [10]. The financial regulators want to see a transparent explainable models and restrict the use of black-box techniques. The new solutions are explainable AI, hybrid modeling, and federated learning. Research should be conducted in the future on reliable AI, reporting of regulations in real time, and interconnection with central bank monitoring. Table 17.4 explains the advantages of AI-based risk systems compared to conventional methods.

TABLE 17.4: Traditional vs. AI-Based Risk Management Systems

Capability	Traditional	AI-Based
Forecasting	RuleBased	DataDriven
StressAdaptation	Static	Dynamic
RealtimeAnalysis	Limited	Continuous
Explainability	High	Moderate
Scalability	Low	High
RiskDetection	Manual	Automated
ScenarioGeneration	Historical	Simulated
DecisionSupport	Expert	Intelligent
LearningAbility	Fixed	Adaptive
AnomalyDetection	Threshold	Predictive
SystemIntegration	Fragmented	Unified
ResponseSpeed	Slow	Rapid
ModelUpdating	Periodic	Continuous
DataProcessing	Batch	Streaming
GovernanceSupport	Procedural	Embedded
Personalization	Generic	Contextual
ResourceAllocation	Heuristic	Optimized
RiskVisualization	Static	Interactive
AuditSupport	Manual	Automated
ComplianceTracking	Reactive	Proactive
ErrorHandling	Corrective	Preventive
KnowledgeManagement	Documented	Intelligent
DecisionConsistency	Variable	Stable
CostEfficiency	Moderate	Optimized
SystemResilience	Limited	SelfHealing

17.5 Conclusion

The chapter has conducted a review of AI-based risk modeling and financial services stress testing, focusing on sophisticated learning methods, generation of scenarios, and deployment aspects. AI-based models are characterized by high levels of accuracy, flexibility, and risk awareness on a system-wide level. The development that will come in the future must be focused on transparency, strength and regulatory alignment. By encompassing interdisciplinary research and technological advancement, AI will remain among the main stakeholders in strengthening financial resiliency and stability.

Bibliography

[1] Diana Ailyn. Feature engineering for financial market prediction: From historical data to actionable insights. *Journal of Financial Data Science*, 6(4):112–128, 2024.

[2] Basel Al-Mansour. Dynamic stress testing and ai: Assessing bank resilience under extreme economic scenarios. *Journal of Risk and Financial Management*, 18(2):88, 2025.

[3] Anonymous. Bayesian modeling for uncertainty management in financial risk forecasting and compliance, 2025.

[4] Thorsten Beck, Vasso Ioannidou, Enrico C. Perotti, Antonio Sánchez Serrano, Javier Suarez, and Xavier Vives. Addressing banks' vulnerability to deposit runs: Revisiting the facts, arguments and policy options. *European Systemic Risk Board Working Paper Series*, (148), 2024.

[5] S. Deng and L. Wu. Adaptive learning in financial risk analytics: High-dimensional data and scenario-based forecasting. *Journal of Financial Data Science*, 7(1):45–62, 2025.

[6] Antonio Díaz, Tjeerd M Boonman, and Abayomi Kupabado. Volatility modeling and tail risk estimation of financial assets: Evidence from gold, oil, bitcoin, and stocks for selected markets. *Risks*, 13(1):12, 2025.

[7] European Central Bank. 2025 stress test: Assessing bank resilience to geopolitical and macro-financial shocks. Technical report, ECB, 2025.

[8] Vijay Govindarajan, Pratik Patel, Sahil Tripathi, Md Azizul Hoque, and Gautam Siddharth Kashyap. Magic-enhanced keyword prompting for zero-shot audio captioning with clip models. *arXiv preprint arXiv:2509.12591*, 2025.

[9] International Monetary Fund. Global financial stability report: Shifting ground beneath the calm. Technical report, IMF, October 2025.

[10] Nir Kshetri. The role of artificial intelligence in financial cybersecurity. *Computer*, 57(2):104–109, 2024.

[11] Dechawat Likitratcharoen and Nalinee Suwannamalik. Assessing financial stability in turbulent times: A study of garch-type value-at-risk model performance. *Risks*, 12(3):51, 2024.

[12] H. Lin, J. Wang, and et al. Trustworthy ai in finance: Balancing model interpretability and predictive power. *Expert Systems with Applications*, 262:125610, 2025.

[13] S. Nadeem. Ai-driven forecasting vs. traditional econometric models: A comparative analysis of predictive accuracy. *Advance Journal of Econometrics and Finance*, 2(1):45–58, 2024.

[14] Ashimiyu Nafiu. Risk management strategies: Navigating volatility in complex financial market environments. *World Journal of Advanced Research and Reviews*, 25(1):236–250, 2025.

[15] Ahmed Osman and Nadia El-Gendy. Big data in financial risk management: Evidence, advances, and open questions: A systematic review. *Frontiers in Artificial Intelligence*, 8, 2025.

[16] Antonio Serrano, Tjeerd Boonman, and Abayomi Kupabado. Generative ai for financial stress testing: A multi-agent simulation approach. *Journal of Risk and Financial Management*, 18(1):42, 2025.

[17] Arpita Soni, Sahil Tripathi, Gautam Siddharth Kashyap, Manaswi Kulahara, Mohammad Anas Azeez, Zohaib Hasan Siddiqui, Nipun Joshi, and Jiechao Gao. Can we predict your next move without breaking your privacy? *arXiv preprint arXiv:2507.08843*, 2025.

[18] Y. Weng. Reinforcement learning in financial decision making: A systematic review of performance and implementation. *Research in International Business and Finance*, 71:102450, 2025.

[19] J. Zheng, R. Miller, and K. Tan. Early warning systems for financial crises: Integrating transformers and graph neural networks. *Finance Research Letters*, 62:105123, 2025.

[20] Y. Zhu, S. I. Taasim, and A. Daud. Data quality and integrity in ai-driven financial risk prediction: A systematic evaluation. *Systems and Soft Computing*, 7:100084, 2025.

[21] Yilin Zhu, Shairil Izwan Taasim, and Adrian Daud. Volatility modeling and tail risk estimation of financial assets: Evidence from gold, oil, bitcoin, and stocks for selected markets. *Risks*, 13(7):138, 2025.

Chapter 18

Decision Intelligence in Healthcare Finance and Insurance Systems

Zohaib Hasan Siddiqui

18.1 Introduction

Healthcare finance [6] and insurance systems [11] operate in complex environments which is characterized by growing medical costs, regulatory constraints, demographic changes, and uncertainty in patient outcomes [17]. Decision intelligence uses artificial intelligence [20], analytics [18] and domain knowledge [1] to support informed financial and operational decision-making in these systems.

There are some traditional decision-support tools which mainly trusts heavily on manual analysis [3] and rule-based methods [4], which struggle to scale with increasing data volume and complexity. Recent advances in machine learning [22], big data analytics [7] and intelligent optimization [14] have enabled automated risk assessment, cost prediction, fraud detection and personalized insurance planning. Table 18.1 shows an overview of key decision intelligence approaches used in healthcare finance and insurance systems.

This chapter surveys havin methodologies in Section 18.2, examines practical applications in Section 18.3, discusses deployment challenges in Section 18.4 and concludes with future research directions in Section 18.5.

18.2 AI-Based Modeling for Healthcare Finance and Insurance

AI-based decision intelligence models which uses heterogeneous data sources [13], including electronic health records [19], claims databases [20], demographic profiles [5] and market indicators [20]. These models mainly enable predictive analytics for medical expenditure, insurance premiums and

DOI: 10.1201/9781003774815-18

TABLE 18.1: AI Models for Financial Decision Support in Healthcare

ModelType	UseCase	KeyStrength
LinearModels	CostEstimation	Simplicity
RandomForest	RiskScoring	Robustness
NeuralNetworks	ClaimsAnalysis	Nonlinearity
GradientBoosting	PremiumPrediction	Precision
ReinforcementLearning	PolicyDesign	Adaptivity
SupportVectorMachines	FraudClassification	Margin
NaiveBayes	ClaimFiltering	Efficiency
DecisionTrees	CostAnalysis	Interpretability
KNearestNeighbors	PatternMatching	Similarity
Autoencoders	AnomalyDetection	Compression
TransformerModels	TextAnalysis	Context
GraphNeuralNetworks	RelationModeling	Connectivity
ExtremeLearningMachines	RiskForecasting	Speed
DeepBeliefNetworks	FeatureLearning	Representation
ConvolutionalNetworks	ImageAnalysis	Extraction
RecurrentNetworks	SequenceModeling	Memory
CapsuleNetworks	StructureLearning	Hierarchy
BayesianNetworks	UncertaintyModeling	Inference
SparseModels	FeatureSelection	Efficiency
MixtureModels	ClusterAnalysis	Flexibility
TransferLearning	KnowledgeReuse	Generalization
SelfSupervisedModels	RepresentationLearning	Scalability
FewShotModels	RapidAdaptation	Efficiency
AttentionModels	SignalPrioritization	Focus
FederatedModels	DistributedLearning	Privacy

reimbursement strategies. Some Supervised learning models that support disease cost prediction [15] and patient risk profiling [10] while unsupervised techniques identify abnormal billing patterns [8]. Reinforcement learning is increasingly used for adaptive pricing and policy optimization [12]. Table 18.2 summarizes commonly used AI models for healthcare financial decision-making.

Feature engineering, dimensionality reduction, and model calibration play important roles in securing reliable performance. Using clinical and financial indicators improves decision relevance and policy performance.

18.3 Applications in Healthcare Insurance and Financial Management

Decision intelligence supports multiple types of operational and strategic functions in healthcare finance and insurance. These include fraud detection [21],

TABLE 18.2: AI Models for Financial Decision Support in Healthcare

ModelType	UseCase	KeyStrength
LinearModels	CostEstimation	Simplicity
RandomForest	RiskScoring	Robustness
NeuralNetworks	ClaimsAnalysis	Nonlinearity
GradientBoosting	PremiumPrediction	Precision
ReinforcementLearning	PolicyDesign	Adaptivity
SupportVectorMachines	FraudClassification	Margin
NaiveBayes	ClaimFiltering	Efficiency
DecisionTrees	CostAnalysis	Interpretability
KNearestNeighbors	PatternMatching	Similarity
Autoencoders	AnomalyDetection	Compression
TransformerModels	TextAnalysis	Context
GraphNeuralNetworks	RelationModeling	Connectivity
ExtremeLearningMachines	RiskForecasting	Speed
DeepBeliefNetworks	FeatureLearning	Representation
ConvolutionalNetworks	ImageAnalysis	Extraction
RecurrentNetworks	SequenceModeling	Memory
CapsuleNetworks	StructureLearning	Hierarchy
BayesianNetworks	UncertaintyModeling	Inference
SparseModels	FeatureSelection	Efficiency
MixtureModels	ClusterAnalysis	Flexibility
TransferLearning	KnowledgeReuse	Generalization
SelfSupervisedModels	RepresentationLearning	Scalability
FewShotModels	RapidAdaptation	Efficiency
AttentionModels	SignalPrioritization	Focus
FederatedModels	DistributedLearning	Privacy

dynamic pricing [9], reimbursement planning [2], patient segmentation [9] and investment management [2]. There are some AI-driven systems that enable insurers to detect abnormal claim patterns, hospitals to optimize revenue cycles and regulators to monitor systemic risks. Personalized insurance plans which can be generated based on patient risk profiles and treatment histories. Table 18.3 presents major application domains of decision intelligence in healthcare finance.

The use of predictive and prescriptive analytics enables proactive financial planning and improved affordability of the patient.

18.4 Deployment Challenges and Future Directions

Despite all the technological progress deploying decision intelligence systems in healthcare finance presents imoortant challenges [9]. There are some issues

TABLE 18.3: Applications of Decision Intelligence in Healthcare Finance

Application	Stakeholders	Outcome
FraudDetection	Insurers	Savings
RiskManagement	Providers	Control
PremiumOptimization	Designers	Fairness
RevenueManagement	Hospitals	Stability
InvestmentPlanning	Administrators	Balance
ClaimsProcessing	Adjusters	Efficiency
ResourceAllocation	Managers	Optimization
CostForecasting	Analysts	Accuracy
PatientBilling	Patients	Transparency
PolicyEvaluation	Regulators	Compliance
MarketAnalysis	Executives	Insight
UtilizationReview	Auditors	Accountability
DemandPrediction	Planners	Preparedness
VendorSelection	Procurement	Reliability
RiskScoring	Underwriters	Assessment
PricingStrategy	Actuaries	Competitiveness
CashFlowModeling	FinanceTeams	Liquidity
ContractManagement	LegalTeams	Consistency
BenefitDesign	Policymakers	Equity
CapacityPlanning	Operators	Readiness
FraudPrevention	Investigators	Protection
QualityAssessment	Inspectors	Improvement
PortfolioOptimization	Investors	Return
ReimbursementAnalysis	Payers	Alignment
ComplianceMonitoring	Officers	Assurance

like data privacy regulations, interoperability issues, model interpretability requirements and organizational resistance hinder large-scale adoption [16]. Bias in training data may lead to unfair pricing or coverage exclusion [23] while limited explainability complicates regulatory approval. Ensuring cybersecurity and maintaining data integrity are additional types of concerns. Table 18.4 summarizes major deployment challenges and corresponding mitigation strategies.

Future research should focus on trustworthy AI, regulatory-compliant architectures and human-in-the-loop decision systems to increase system reliability and acceptance.

18.5 Conclusion

This chapter has reviewed decision intelligence frameworks for healthcare finance and insurance systems by showing AI-based modeling techniques, application domains and deployment challenges. Survey-style comparative

TABLE 18.4: Deployment Challenges and Mitigation Strategies

Challenge	Impact	Mitigation
DataPrivacy	LegalRisk	FederatedLearning
ModelTransparency	LowTrust	ExplainableAI
SystemIntegration	OperationalDelay	APIStandardization
BiasFairness	EthicalIssues	FairnessAuditing
Scalability	PerformanceLimits	EdgeCloud
Latency	Delays	Caching
Security	Breaches	Encryption
Reliability	Downtime	Redundancy
Maintainability	Complexity	Automation
Interoperability	Fragmentation	Standardization
Compliance	Penalties	Governance
DataQuality	Inaccuracy	Validation
ModelDrift	Degradation	Retraining
UserAdoption	Resistance	Training
CostManagement	Overspending	Optimization
ExplainabilityGap	Misinterpretation	Visualization
ResourceConstraints	Bottlenecks	Scheduling
VendorLockin	Dependency	Portability
SkillShortage	Inefficiency	Upskilling
TestingCoverage	Failures	Automation
ConfigurationErrors	Instability	Verification
PrivacyLeakage	Exposure	Anonymization
ModelVersioning	Confusion	Tracking
IncidentResponse	Escalation	Playbooks
PerformanceMonitoring	Blindspots	Telemetry

analyses has showed the growing role of intelligent analytics in financial governance and policy design. Advances in explainable AI, privacy-preserving learning and hybrid intelligence are expected to further improve decision quality and institutional resilience. Through aligning technological innovation with ethical and regulatory requirements, decision intelligence will remain central to sustainable healthcare financial management.

Bibliography

[1] Asmae ABADI, Chaimae ABADI, and Mohammed ABADI. Intelligent decision making and knowledge management system for industry 4.0 maturity assessment. *Statistics, Optimization & Information Computing*, 14(1):207–228, 2025.

[2] Yusuff Taofeek Adeshina. Strategic implementation of predictive analytics and business intelligence for value-based healthcare performance optimization in us health sector.

[3] Shania Karim Adittee. Automated decision support systems in innovation: the role of decision support systems in marketing and product development. 2025.

[4] Ashraf Alnattah, Mahdie Jajroudi, Seyyed Ali N Fadafen, Mahdi N Manzari, Saeid Eslami, Seyyed Ali Najafi Fadafen, and Mahdi Najafi Manzari. Artificial intelligence in clinical decision-making: a scoping review of rule-based systems and their applications in medicine. *Cureus*, 17(8), 2025.

[5] Aditya Dhanekula. Ai-driven business intelligence framework for predictive decision-making and strategic resource optimization. *International Journal of Business and Economics Insights*, 5(3):1238–1270, 2025.

[6] Cleopatra Elshiekh, Roberta Rudà, Edward R Scheffer Cliff, Francesca Gany, and Joshua A Budhu. Financial challenges of being on long-term, high-cost medications. *Neuro-oncology practice*, 12(Supplement_1):i49–i58, 2025.

[7] Vijay Govindarajan, Pratik Patel, Sahil Tripathi, Md Azizul Hoque, and Gautam Siddharth Kashyap. Magic-enhanced keyword prompting for zero-shot audio captioning with clip models. *arXiv preprint arXiv:2509.12591*, 2025.

[8] Mohammad Hassan and Mahmud Alam. Automating healthcare claims processing with supervised and unsupervised ai models. *Journal of Applied Big Data Analytics, Decision-Making, and Predictive Modelling Systems*, 9(2):1–11, 2025.

[9] Shankar Subramanian Iyer. Data-driven decision making: The key to future health care business success. 2025.

[10] Bolatito khadijat Aremu and Olagunju Omotola Peggy. Advanced machine learning-driven business analytics for real-time health risk stratification and cost prediction models. *World J. Adv. Res. Rev*, 26(2):150–167, 2025.

[11] Venkata Sambasivarao Kopparapu. Healthcare insurance data infrastructure: A comprehensive analysis of edi standards and processing systems. *International Journal of Research in Computer Applications and Information Technology (IJRCAIT)*, 8(1):2341–2353, 2025.

[12] Fabian Lange, Leonard Dreessen, and Rainer Schlosser. Reinforcement learning versus data-driven dynamic programming: a comparison for finite horizon dynamic pricing markets: F. lange et al. *Journal of Revenue and Pricing Management*, pages 1–17, 2025.

[13] Md Nur Hasan Mamun. Role of ai and data science in data-driven decisionmaking for it business intelligence: A systematic literature review. *Available at SSRN 5402976*, 2025.

[14] Courage Oko-Odion. Ai-driven risk assessment models for financial markets: Enhancing predictive accuracy and fraud detection. *International Journal of Computer Applications Technology and Research*, 14(04):80–96, 2025.

[15] Santhosh Kumar Pendyala. Healthcare value-based reimbursement: A predictive analytics and machine learning framework for cost optimization and quality improvement. 2025.

[16] Rvs Praveen, Anurag Shrivastava, Gunjan Sharma, Aboothar Mahmood Shakir, Manish Gupta, and Satya Subrahmanya Sai Ram Gopal Peri. Overcoming adoption barriers strategies for scalable ai transformation in enterprises. In *2025 International Conference on Engineering, Technology & Management (ICETM)*, pages 1–6. IEEE, 2025.

[17] Imran Qureshi and Naveed Hassan. Experimental analysis of health insurance market reforms and their effects on financial risk mitigation and economic resilience in vulnerable groups. *International Review of Experimental Sciences, Scientific Discoveries, and Technological Advancements*, 9(5):1–9, 2025.

[18] Muslima Begom Riipa, Nazma Begum, Md Sakib Hasan Hriday, and Syed Azazul Haque. Role of data analytics in enhancing business decision-making and operational efficiency. *International Journal of Communication Networks and Information Security*, 17(2):400–412, 2025.

[19] Ganesh Shankar Sargam. Ai-driven clinical decision support: Enhancing care quality through ehr-integrated intelligence. *International Journal of Advanced Research in Computer Science*, 16(3), 2025.

[20] Hozyfa Shafa. Artificial intelligence-driven business intelligence models for enhancing decision-making in us enterprises. *ASRC Procedia: Global Perspectives in Science and Scholarship*, 1(01):771–800, 2025.

[21] Arpita Soni, Sahil Tripathi, Gautam Siddharth Kashyap, Manaswi Kulahara, Mohammad Anas Azeez, Zohaib Hasan Siddiqui, Nipun Joshi, and Jiechao Gao. Can we predict your next move without breaking your privacy? *arXiv preprint arXiv:2507.08843*, 2025.

[22] Naga Harsha Tumma. Machine learning in finance and industry to address risk management, automation, and personalized services.

[23] Shenggang Yang, Wenbo Ma, Sangzhi Zhu, Ao Shu, and Yue Zhou. Data bias, algorithmic discrimination and the fairness issues of individual credit accessibility. *Journal of Accounting Literature*, 2025.

Part V

AI-Enhanced Decision-Making in Business and Finance

Chapter 19

Unleashing Human Potential: Advancing Cognitive Capabilities Through AI-Designed Neural Interfaces

Venu Madhav Aragani

19.1 Introduction

The most complex organ that science has ever heard of is the human brain, which has a complex network of billions of neurones that makes it up [10]. The human brain is a complex network of billions of neurones and it is the most complex organ to be discovered by science. It is the origin of our memories, thoughts, feelings, and consciousness and hence it comprises the foundation of our cognitive abilities. The enhancement of memory, learning, and sensory perception has been a topic of interest since ancient times that have resulted in the emergence of many methods and technology in this field [12]. The emergence of Artificial Intelligence (AI) in recent years [22] as well as coming up with brain interfaces, has presented new opportunities in this pursuit based on the potential to significantly improve the cognitive capabilities of humans [35] as a goal that was once unattainable.

Another term used to refer to neural interfaces is the Brain-Computer Interfaces (BCIs) [34] which enables a direct brain-to-external device interface. They can be used with healthy people to enhance cognitive functions and also in repairing lost motor or sensory functions. It can be said that AI will greatly improve the neural interface capabilities, enabling a more accurate interpretation [43] of brain signals, adaptive learning, and personalised user experiences [38] which are present in the future [37].

The neural interfaces developed by AI have an extensive impact on the life of people and the society in general. Although these technologies present valuable ethical and social concerns, they can result in healthcare [32], education [19], and worker productivity [1] breakthroughs and increased productivity of workers [1]. In an attempt to get a clear picture of the present position of

DOI: 10.1201/9781003774815-19

TABLE 19.1: Neural Signal Acquisition and Processing Techniques

Method	**Advantage**	**Limitation**
EEG	Noninvasive	Noise
ECoG	Resolution	Surgery
Microelectrodes	Precision	Invasiveness
fNIRS	Portability	Depth
MEG	Sensitivity	Cost
LFP	Stability	Localization
Optogenetics	Specificity	Genetics
Wearables	Mobility	Accuracy
Implants	Reliability	Infection

the neural interfaces designed by AI, this chapter will examine the challenges that are yet to be addressed and the technological breakthroughs that have brought us this far. The moral concerns that should be addressed during the development and utilization of novel technologies include privacy, consent, and equity, which we will discuss. The possible long-term social impacts of AI-designed neural interfaces will also be argued out with special focus on both the benefits and drawbacks.

19.2 AI-Designed Neural Interfaces: Current State and Technological Foundations

A significant advancement in AI [36], engineering [8], and neuroscience [2] is the creation of brain interfaces created by AI. Direct brain-to-external device connection is made possible by these interfaces, which have developed from simple systems into highly advanced instruments that can accurately read complicated neurological signals. This section examines the present status of AI-designed neural interfaces and the technological underpinnings that support these developments, such as neural signal processing [14], acquisition [16], and AI algorithm integration [26].

19.2.1 Neural Signal Acquisition and Processing

Effectively acquiring and processing brain signals is the fundamental function of any neural interface (see Table 19.1). The process of acquiring neural signals entails recording the electrical activity of neurones. There are several ways to accomplish this, each with pros and cons. Electroencephalography (EEG), intracortical microelectrodes, and electrocorticography (ECoG) are the most widely used techniques [33].

TABLE 19.2: AI Algorithms for Neural Signal Decoding

Algorithm	Strength	Application
SVM	Classification	Motor
RandomForest	Robustness	Diagnosis
CNN	Featurelearning	Recognition
RNN	Sequencing	Prediction
LSTM	Memory	Control
Transformer	Attention	Decoding
Autoencoder	Compression	Denoising
ReinforcementLearning	Adaptation	Optimization
Bayesian	Uncertainty	Modeling

- Electroencephalography (EEG) [9]: EEG is a non-invasive method that records electrical activity along the scalp by showing the synchronized activity of large groups of neurons. While EEG is mainly used due to its non-invasive nature it suffers from relatively low spatial resolution and is susceptible to noise and interference from external type of sources.
- Intracortical Microelectrodes [42]: These electrodes are implanted directly into the brain's cortex by allowing for the recording of neural activity at the level of individual neurons or small groups of neurons. There are some microelectrodes that provide high spatial and temporal resolution by making them ideal for precise control in applications like prosthetic devices. They are highly invasive and give risks like tissue damage and infection.
- Electrocorticography (ECoG) [17]: ECoG includes placing electrodes on the surface of the brain, having the skull but outside the brain tissue. This method gives a balance between invasiveness and signal quality with better type of spatial resolution than EEG and lower risks than microelectrodes. ECoG is mainly used in clinical settings for epilepsy monitoring and has shown promise in BCI applications.

19.2.2 AI Algorithms in Neural Interfaces

The use of AI algorithms into neural interfaces has been a game-changer (see Table 19.2) by enabling more accurate interpretation of neural signals [29] and more responsive control systems [6]. AI algorithms mainly those based on machine learning [43] and deep learning [30] which are quite well-suited to handling the complexity and variability of neural data.

- Machine Learning Algorithms: There are some traditional machine learning algorithms like support vector machines (SVMs) [15] and random forests [20] which have been used to classify neural signals with different degrees of success. These algorithms require careful feature selection and engineering which can be quite challenging given the high

TABLE 19.3: Integration of Neural Interfaces with External Systems

System	Function	Benefit
Prosthetics	Movement	Independence
BCI	Communication	Accessibility
Wheelchairs	Navigation	Mobility
Exoskeletons	Assistance	Strength
Robotics	Manipulation	Precision
Computers	Interaction	Productivity
SmartHomes	Automation	Comfort
ARVR	Immersion	Engagement
Implants	Feedback	Sensation

dimensionality and noise inherent in neural data. They have proven good in applications where the neural signals are relatively well-understood like motor control.

- Deep Learning Algorithms: Deep learning which is a subset of machine learning that has shown tremendous promise in neural interfaces due to its ability to automatically learn features from raw data. Convolutional Neural Networks (CNNs) [41], Recurrent Neural Networks (RNNs) [27] and their variants have been successfully applied to neural signal classification by enabling more strong and accurate decoding.
- Reinforcement Learning: Reinforcement learning (RL) [4] algorithms are also being explored in neural interfaces mainly for adaptive control systems. RL allows a system to learn optimal control strategies through trial and error by making it good for applications where the environment or the user's needs may change over time.

19.2.3 Integration of Neural Interfaces with External Devices

Neural interfaces' ultimate goal is to provide smooth use with external equipment so that the brain of the user can control computers (see Table 19.3), prosthetic limbs and other technology [21]. This gives both precise neural signal decoding and powerful external device connectivity.

- Prosthetic Devices: Neural interfaces have been effectively combined with the use of prosthetic limbs, whereby people who had amputation use their thoughts to make the prosthetic limbs behave. Artificial intelligence algorithms are useful in neural signal to movement [5] conversion to more natural, intuitive control with AI algorithms enabling more precise movements in robots and other computer-controlled devices like drones or robotic nurses. Such studies [18] have demonstrated that users can acquire the ability to control the prosthetics limbs with high level

TABLE 19.4: Challenges in AI-Designed Neural Interfaces

Challenge	Cause	Impact
Noise	Artifacts	Inaccuracy
Latency	Computation	Delay
Overload	Bigdata	Bottleneck
Biocompatibility	Immunity	Degradation
Scalability	Complexity	Limitation
Privacy	Surveillance	Misuse
Ethics	Consent	Conflict
Security	Hacking	Breach
Durability	Wear	Failure

of accuracy and the current developments are working on further developing the sensory feedback in order to make the experience of the user more advanced.

- Brain-Computer Interfaces (BCIs) to Communicate: BCIs to communicate enable people with the most severe motor impairments, including those with Amyotrophic Lateral Sclerosis (ALS), to communicate through their brain signals by controlling a computer cursor or typing on a virtual keyboard to communicate with a computer screen [40]. The decoding of the intentions of the user in the neural activity is impossible without AI algorithms that make it possible to communicate with the computer in real-time in interaction with it Without AI algorithms, the user cannot be involved in interaction with the computer in real-time, and it is impossible to decode the intentions of the user in the neural activity of the user and others [24]. The development of AI has greatly contributed to the speed and accuracy of such systems making them more practical in their daily application.
- Augmented and Virtual Reality (AR/VR) Integration: The neural interface connecting to AR and VR systems is a promising possibility of integration into the future [23]. By integrating neural control with immersive environments, the systems can bring up completely new ways of interaction and experience.

19.2.4 Current Challenges and Limitations

Although AI-based neural interfaces have been developed to remarkable progress, there are still various challenges (see Table 19.4). The issue of signal accuracy and reliability is one of the major problems. Even the most sophisticated AI algorithms can find it difficult to decode their neural signals [3], and this is especially true when internal noise and artifacts are prominent and external noise and artifacts are involved in the real-life environment.

- Data Overload: The amount of data produced by the interfaces of neural interfaces is enormous thus presenting a great challenge in storage, processing as well as analysis. The AI algorithms should be able to process such data effectively and at the same time maintain high degree of accuracy. Moreover, real-time processing may be needed and this may be computationally challenging.
- Biocompatibility and Longevity: In the case of invasive neural interface biocompatibility is a major issue. The immune system of the body against the implanted devices may cause degradation of signals with time, which decreases the efficiency of the interface. Studies are in progress to make materials and methods that will reduce the immune response and increase the life span of such devices.
- Ethical and Privacy Concerns: AI in neural interfaces puts a lot of ethical and privacy issues into the limelight. Possessing the capabilities of accessing and manipulating neural data opens the possibility of misuse, including the possible violations of personal thinking or the establishment of cognitive biases. It is essential that these technologies are created and deployed in a manner that would not violate the privacy and autonomy of individuals.

19.3 Enhancing Cognitive Capabilities Through Neural Interfaces

19.3.1 Mechanisms of Cognitive Enhancement

There are some neural interfaces that enable cognitive enhancement by establishing a direct connection between the brain and external devices like computers or robotic systems [13] (see Table 19.5). These interfaces can

TABLE 19.5: Mechanisms of Cognitive Enhancement via Neural Interfaces

Stage	Function	Outcome
Acquisition	Recording	Digitization
Filtering	Cleaning	Clarity
Decoding	Translation	Commands
Stimulation	Activation	Modulation
Feedback	Adjustment	Optimization
Adaptation	Learning	Stability
Synchronization	Alignment	Accuracy
Calibration	Tuning	Precision
Validation	Testing	Reliability

TABLE 19.6: Neural Interface Applications for Memory Enhancement

Technique	Target	Benefit
Stimulation	Hippocampus	Retention
Encoding	Cortex	Recall
Reinforcement	Synapses	Consolidation
Monitoring	Activity	Optimization
Modulation	Networks	Stability
Recording	Patterns	Retrieval
Reactivation	Circuits	Recovery
Feedback	Signals	Adaptation
Calibration	Parameters	Accuracy

either stimulate neural activity or decode neural for signals to interact with other systems. The process generally includes several steps:

1. Signal Acquisition: Signal acquisition uses electrodes or sensors to record neural signals that are digitized to process them.
2. Signal Processing and Interpretation: AI algorithms filter and interpret such signals, which are then translated into commands or responses. Machine learning methods, especially state-of-the-art methods, are especially well-suited to find complicated trends in neural data, and this can be used to predict or manipulate mental states.
3. Feedback Loop: The interface may have a feedback mechanism, which can be adjusted in real-time in response to the reaction of the brain to stimulation. This is also an important adaptive learning process that enhances the efficacy of the neural interfaces.

19.3.2 Applications in Memory Enhancement

One of the most promising areas for neural interface-based cognitive enhancement is memory [39] (see Table 19.6). Scientists have developed devices that improve memory recall and retention by focusing on particular brain regions, such as the hippocampus [7].

19.3.3 Accelerating Learning

Additionally, neuronal interfaces can quicken the learning process by improving synaptic plasticity the brain's capacity to fortify connections between neurons (see Table 19.7). When the brain is in a condition favourable to learning, AI-driven BCIs can identify it and provide targeted stimulation to strengthen the neural networks linked to particular tasks or knowledge. This could enable people to pick up new abilities more quickly and has significant ramifications for education and skill development [43].

TABLE 19.7: Neural Interface Techniques for Learning Acceleration

Method	Process	Effect
Stimulation	Plasticity	Strengthening
Detection	Readiness	Timing
Optimization	Networks	Efficiency
Reinforcement	Pathways	Retention
Adaptation	Feedback	Improvement
Synchronization	Activity	Coherence
Prediction	Patterns	Anticipation
Calibration	Signals	Precision
Personalization	Profiles	Customization

TABLE 19.8: Sensory Augmentation Using Neural Interfaces

Sense	Technology	Outcome
Hearing	Cochlear	Perception
Vision	Retinal	Restoration
Touch	Haptics	Sensation
Balance	Vestibular	Stability
Smell	Olfactory	Detection
Taste	Gustatory	Recognition
Proprioception	Feedback	Awareness
Temperature	Thermosensory	Regulation
Pressure	Tactile	Sensitivity

19.3.4 Sensory Augmentation

Using neural interfaces, sensory augmentation aims to improve or restore sensory experiences (see Table 19.8). For those who are deaf, cochlear implants are a well-known type of sensory augmentation [28]. Recent developments have expanded this idea to include touch and vision [25].

19.4 Challenges and Ethical Considerations

19.4.1 Technical Challenges

Despite the good applications there are still some technical challenges that hides the great use of AI-designed neural interfaces. These challenges include (see Table 19.9):

TABLE 19.9: Technical Challenges in AI-Designed Neural Interfaces

Challenge	Cause	Impact
Precision	Noise	Inaccuracy
Overload	Volume	Complexity
Latency	Processing	Delay
Scalability	Resources	Limitation
Stability	Drift	Degradation
Integration	Compatibility	Rejection
Durability	Wear	Failure
Calibration	Variability	Misalignment
Sensitivity	Interference	Distortion

- Signal Precision: It is still challenging to have a fine control over neural signals, especially in non-invasive systems where the quality of signals is commonly impaired.
- Data Overload: The huge data produced by the neural interfaces needs to be processed by specialized AI algorithms in order to be effectively interpreted. Nevertheless, the process of deriving valuable insights out of such data is complicated and taxing regardless of the advanced machine learning method.
- Interaction with Biological Systems: This is still a challenge to ensure that neural interfaces can be easily integrated into biological systems without causing any unwanted reactions such as inflammation or permanent damage.

19.4.2 Ethical Issues

The ethical implications of cognitive enhancement through neural interfaces are profound. Key concerns include (see Table 19.10):

- Privacy and Consent: The ability of neural interfaces to access and modify cognitive functions raises significant privacy concerns. Ensuring that individuals have full control over their neural data and providing informed consent are critical ethical considerations.
- Equity and Access: The development and deployment of neural interfaces could exacerbate existing inequalities if access to these technologies is limited to those who can afford them. This could lead to a future where cognitive enhancement is a privilege of the wealthy, further widening social divides [31].

TABLE 19.10: Ethical Issues in Cognitive Enhancement Technologies

Issue	Concern	Risk
Privacy	Surveillance	Exposure
Consent	Autonomy	Coercion
Equity	Access	Inequality
Ownership	Data	Exploitation
Manipulation	Control	Dependency
Bias	Algorithms	Discrimination
Transparency	Opacity	Mistrust
Misuse	Abuse	Harm
Identity	Cognition	Alienation

TABLE 19.11: Regulatory Frameworks for AI-Based Neural Interfaces

Domain	Focus	Objective
Privacy	Protection	Security
Safety	Certification	Reliability
Ethics	Compliance	Responsibility
Standards	Harmonization	Consistency
Governance	Oversight	Accountability
Licensing	Approval	Control
Audit	Monitoring	Transparency
Liability	Responsibility	Justice
Collaboration	Alignment	Interoperability

- Potential for Misuse: There is a risk that neural interfaces could be used for purposes beyond cognitive enhancement, such as surveillance or coercion. Safeguards need to be in place to prevent such misuse and to ensure that these technologies are used ethically and responsibly.

19.4.3 Regulatory Frameworks

The legal regulations that apply to the AI-based neural interfaces also need to transform with the field (see Table 19.11). The existing rules are often not in line with the technological changes, and there are always holes that might be used. The specific barriers of the neural interfaces, including protection of personal freedom and morality of the application of these tools, must be addressed with a comprehensive regulatory approach in order to meet the challenge [11]. It will be necessary to create an effective regulatory system through international cooperation and the standardization process.

19.5 Conclusion

The neural interfaces created by AI represent a very new breakthrough in the pursuit of improving human cognitive abilities. Advanced AI algorithms can drive these interfaces to use in enhancing memory, learning, and sensory augmentation in a significant way. Nevertheless, the creation and implementation of these technologies are faced with great challenges, such as the technical constraints, the ethical issues, and the necessity of effective regulatory mechanisms.

The future of neural interfaces is promising and it has the capability to change the life of not only an individual, but also the whole society. As we keep stretching the limits of our possibilities, it is important to consider the consequences of what we are doing carefully and seek ways of using these technologies to the greatest benefit and the least harm.

Bibliography

[1] Muhammad Abbas Khan, Habib Khan, Muhammad Faizan Omer, Inam Ullah, and Muhammad Yasir. Impact of artificial intelligence on the global economy and technology advancements. In *Artificial General Intelligence (AGI) Security: Smart Applications and Sustainable Technologies*, pages 147–180. Springer, 2024.

[2] Zarif Bin Akhtar and Victor Stany Rozario. Ai perspectives within computational neuroscience: Eeg integrations and the human brain. In *Artificial Intelligence and Applications*, volume 3, pages 145–160, 2025.

[3] Ali Mohammad Alqudah and Zahra Moussavi. A review of deep learning for biomedical signals: Current applications, advancements, future prospects, interpretation, and challenges. *Computers, Materials & Continua*, 83(3), 2025.

[4] Anuradha M Annaswamy. Adaptive control and intersections with reinforcement learning. *Annual Review of Control, Robotics, and Autonomous Systems*, 6(1):65–93, 2023.

[5] Pedro Garcia Batista, André Costa Vieira, and Pedro Dinis Gaspar. Robotic prostheses and neuromuscular interfaces: A review of design and technological trends. *Machines*, 13(9):804, 2025.

[6] Andrea Calderone, Desirèe Latella, Elvira La Fauci, Roberta Puleo, Arturo Sergi, Mariachiara De Francesco, Maria Mauro, Angela Foti, Leda

Salemi, and Rocco Salvatore Calabrò. Artificial intelligence-driven neuromodulation in neurodegenerative disease: Precision in chaos, learning in loss. *Biomedicines*, 13(9):2118, 2025.

[7] Amalia Călinescu. The impact of digital technologies on memory and memory studies. *Journal of Contemporary Philosophical and Anthropological Studies*, 2(1), 2024.

[8] Zehong Cao. A review of artificial intelligence for eeg-based brain- computer interfaces and applications. *Brain Science Advances*, 6(3):162–170, 2020.

[9] Ahmad Chaddad, Yihang Wu, Reem Kateb, and Ahmed Bouridane. Electroencephalography signal processing: A comprehensive review and analysis of methods and techniques. *Sensors*, 23(14):6434, 2023.

[10] Mike KS Chan, Mohd Iskandar Jumat, Florisa Landa, Nur Shafawati Saili, Siti Azmah Jambo, Dianah Florentius, Nuryasmin Ezzaty, Raz Haziqah Hani Razali, Brenda Son Pei Chui, Aziera Farhanah Adihidayah Suardi, et al. Exploring the complexity of the human brain: Cell types, numbers, and lobar functions. *American Journal of Biomedical Science and Research*, 26(4):445–458, 2025.

[11] Michele Di Salvo. The protection of neural rights in the age of neurotechnologies and ai. the ethical challenge for law and neuroscience. *Russ. J. Econ. & L.*, 19:202, 2025.

[12] Athanasios Drigas and Eleni Mitsea. The 8 pillars of metacognition. *International Journal of Emerging Technologies in Learning (iJET)*, 15(21):162–178, 2020.

[13] A Shaji George. Enhancing human potential: An exploration of spatial computing, polyfunctional robotics, and neural augmentation for human-machine synergy. *Partners Universal Innovative Research Publication*, 3(2):61–73, 2025.

[14] Vijay Govindarajan, Pratik Patel, Sahil Tripathi, Md Azizul Hoque, and Gautam Siddharth Kashyap. Magic-enhanced keyword prompting for zero-shot audio captioning with clip models. *arXiv preprint arXiv:2509.12591*, 2025.

[15] Maria Camila Guerrero, Juan Sebastián Parada, and Helbert Eduardo Espitia. Eeg signal analysis using classification techniques: Logistic regression, artificial neural networks, support vector machines, and convolutional neural networks. *Heliyon*, 7(6), 2021.

[16] Mengyu Huang, Waruna Lakmal Dissanayaka, and Cynthia KY Yiu. Artificial intelligence driven innovation: Advancing mesenchymal stem cell therapies and intelligent biomaterials for regenerative medicine. *Bioengineering*, 12(12):1302, 2025.

[17] Sidrat Tasawoor Kanth and Supratim Ray. Electrocorticogram (ecog) is highly informative in primate visual cortex. *Journal of Neuroscience*, 40(12):2430–2444, 2020.

[18] Mohammadmahdi Karimi, Nashmin Yeganeh, Ivan Makarov, Atli Örn Sverrisson, Karl Fannar Gunnarsson, Kristín Briem, Sigurur Brynjólfsson, Árni Kristjánsson, and Runar Unnthorsson. Haptic feedback systems for lower-limb prosthetic applications: a review of system design, user experience, and clinical insights. *Bioengineering*, 12(9):989, 2025.

[19] Mustafa Kayyali. Future possibilities and challenges of ai in education. In *Transforming education with generative AI: Prompt engineering and synthetic content creation*, pages 118–137. IGI Global Scientific Publishing, 2024.

[20] Souhaila Khalfallah, William Puech, Mehdi Tlija, and Kais Bouallegue. Exploring the effectiveness of machine learning and deep learning techniques for eeg signal classification in neurological disorders. *IEEE Access*, 13:17002–17015, 2025.

[21] Preetam Kumar Khuntia and PV Manivannan. Review of neural interfaces: Means for establishing brain–machine communication. *SN Computer Science*, 4(5):672, 2023.

[22] Kavati Pavan Kumar, PJ Swarubini, and Nagarajan Ganapathy. Cognitive artificial intelligence. In *Artificial Intelligence and Biological Sciences*, pages 301–323. CRC Press, 2025.

[23] Andrea Lastrucci, Yannick Wandael, Angelo Barra, Renzo Ricci, Giovanni Maccioni, Antonia Pirrera, and Daniele Giansanti. Exploring augmented reality integration in diagnostic imaging: Myth or reality? *Diagnostics*, 14(13):1333, 2024.

[24] Manoj Kumar Mahto and G Rajavikram. Fundamentals of ai and communication networks: Applications in human social activities. In *Intelligent Networks*, pages 1–17. CRC Press, 2024.

[25] Albert Manero, Viviana Rivera, Qiushi Fu, Jonathan D Schwartzman, Hannah Prock-Gibbs, Neel Shah, Deep Gandhi, Evan White, Kaitlyn E Crawford, and Melanie J Coathup. Emerging medical technologies and their use in bionic repair and human augmentation. *Bioengineering*, 11(7):695, 2024.

[26] Wubshet Getachew Mengesha. Ai-driven design of multifunctional nanomaterials in revolutionizing high-temperature, high-power solutions for space technology: potentials, challenges and perspectives. *Discover Nano*, 20(1):220, 2025.

[27] Ibomoiye Domor Mienye, Theo G Swart, and George Obaido. Recurrent neural networks: A comprehensive review of architectures, variants, and applications. *Information*, 15(9):517, 2024.

[28] Takahiro Miura et al. Narrative review of assistive technologies and sensory substitution in people with visual and hearing impairment. *Psychologia*, 65(1):70–99, 2023.

[29] Razvan Onciul, Catalina-Ioana Tataru, Adrian Vasile Dumitru, Carla Crivoi, Matei Serban, Razvan-Adrian Covache-Busuioc, Mugurel Petrinel Radoi, and Corneliu Toader. Artificial intelligence and neuroscience: Transformative synergies in brain research and clinical applications. *Journal of Clinical Medicine*, 14(2):550, 2025.

[30] Prajwal Pawar and Punam Shinde. Neuroevolution in artificial intelligence. *algorithms*, 6:10, 2024.

[31] Mihael Techie Quaicoe. Neural interface learning systems: Ethical and pedagogical implications for african higher education by 2040. *International Journal of Humanities, Management, Engineering, Education and Legal Studies*, pages 125–161, 2025.

[32] Soha Rawas. Ai: the future of humanity. *Discover artificial intelligence*, 4(1):25, 2024.

[33] Mehdi Shokoueinejad, Dong-Wook Park, Yei Hwan Jung, Sarah K Brodnick, Joseph Novello, Aaron Dingle, Kyle I Swanson, Dong-Hyun Baek, Aaron J Suminski, Wendell B Lake, et al. Progress in the field of microelectrocorticography. *Micromachines*, 10(1):62, 2019.

[34] Shraddha Jain Shrama and Ratnalata Gupta. Brain–computer interface. In *Machine Learning Models and Architectures for Biomedical Signal Processing*, pages 321–352. Elsevier, 2025.

[35] Bhupinder Singh. Evolutionary global neuroscience for cognition and brain health: Strengthening innovation in brain science. In *Biomedical Research Developments for Improved Healthcare*, pages 246–272. IGI Global Scientific Publishing, 2024.

[36] Arpita Soni, Sahil Tripathi, Gautam Siddharth Kashyap, Manaswi Kulahara, Mohammad Anas Azeez, Zohaib Hasan Siddiqui, Nipun Joshi, and Jiechao Gao. Can we predict your next move without breaking your privacy? *arXiv preprint arXiv:2507.08843*, 2025.

[37] Åsne Stige, Efpraxia D Zamani, Patrick Mikalef, and Yuzhen Zhu. Artificial intelligence (ai) for user experience (ux) design: a systematic literature review and future research agenda. *Information Technology & People*, 37(6):2324–2352, 2024.

[38] Wadim Strielkowski, Veronika Grebennikova, Alexander Lisovskiy, Guzalbegim Rakhimova, and Tatiana Vasileva. Ai-driven adaptive learning for sustainable educational transformation. *Sustainable Development*, 33(2):1921–1947, 2025.

[39] TA Suhail, R Subasree, Vinod AP, et al. On the feasibility of an online brain-computer interface-based neurofeedback game for enhancing attention and working memory in stroke and mild cognitive impairment patients. *Biomedical Physics & Engineering Express*, 11(2):025049, 2025.

[40] Ceci Verbaarschot, Daniëlle Tump, Andreea Lutu, Marzieh Borhanazad, Jordy Thielen, Philip van den Broek, Jason Farquhar, Janneke Weikamp, Joost Raaphorst, Jan T Groothuis, et al. A visual brain-computer interface as communication aid for patients with amyotrophic lateral sclerosis. *Clinical Neurophysiology*, 132(10):2404–2415, 2021.

[41] KE Ch Vidyasagar, K Revanth Kumar, GNK Anantha Sai, Munagala Ruchita, and Manob Jyoti Saikia. Signal to image conversion and convolutional neural networks for physiological signal processing: A review. *Ieee Access*, 12:66726–66764, 2024.

[42] Yang Wang, Xinze Yang, Xiwen Zhang, Yijun Wang, and Weihua Pei. Implantable intracortical microelectrodes: reviewing the present with a focus on the future. *Microsystems & Nanoengineering*, 9(1):7, 2023.

[43] Christopher Williams, Fahim Islam Anik, Md Mehedi Hasan, Juan Rodriguez-Cardenas, Anushka Chowdhury, Shirley Tian, Selena He, and Nazmus Sakib. Advancing brain-computer interface closed-loop systems for neurorehabilitation: Systematic review of ai and machine learning innovations in biomedical engineering. *JMIR Biomedical Engineering*, 10:e72218, 2025.

Chapter 20

Ethical, Legal, and Societal Challenges of AI in High-Stakes Decision Systems

Pushkar Arora

20.1 Introduction

Artificial Intelligence (AI) systems are increasingly deployed in high-stakes decision environments like healthcare [6], finance [23], criminal justice [16], public governance [24] and national security [29]. In these domains, automated decisions may affect individual rights [11], social equity [9] and institutional accountability [3]. While AI gives some important benefits in terms of performance and predictive accuracy its growing influence has raised some ethical, legal and societal concerns. There are some issues which is related to algorithmic bias [7], transparency [26], privacy [25], accountability [2] and human oversight have grown as central challenges in responsible AI deployment. Failures in addressing these concerns may result in discrimination, legal violations, loss of public trust and systemic risks. Table 20.1 provides an overview of major ethical risks which is associated with AI-based decision systems.

This chapter surveys ethical foundations in Section 20.2 which examines legal and regulatory frameworks in Section 20.3, discusses societal impacts and governance in Section 20.4 and concludes with future research directions in Section 20.5.

20.2 Ethical Foundations and Responsible AI Design

Ethical AI design looks to have that intelligent systems align with human values [10], social norms [22] and moral responsibilities [17]. Core ethical principles include fairness, accountability, transparency, beneficence and respect

DOI: 10.1201/9781003774815-20

TABLE 20.1: Major Ethical Risks in High-Stakes AI Systems

Risk	**Domain**	**Impact**
Bias	Recruitment	Discrimination
Opacity	Healthcare	Mistrust
Surveillance	Security	Infringement
Overreliance	Aviation	Dependency
Harm	Welfare	Inequality
Manipulation	Media	Misinformation
Profiling	Marketing	Exploitation
Exclusion	Finance	Marginalization
AutonomyLoss	Robotics	Dependence
DataMisuse	Research	Breach
Misclassification	Justice	Misjudgment
Displacement	Labor	Unemployment
Hallucination	Education	Confusion
Coercion	Governance	Suppression
Monopolization	Industry	Concentration
Disinformation	Politics	Polarization
Censorship	Media	Restriction
Weaponization	Defense	Escalation
Dehumanization	Services	Alienation
Overfitting	Analytics	Instability
Drift	Operations	Degradation
Spoofing	Biometrics	Impersonation
Leakage	Databases	Exposure
Exaggeration	Advertising	Deception
Exploitation	Platforms	Abuse

for autonomy. These principles guide system developers in preventing harmful outcomes and promoting socially beneficial applications [8]. There are some data collection practices like fairness-aware learning methods, bias mitigation techniques [20] and inclusive data collection which are used to reduce discriminatory behavior [21]. There are some explainable AI techniques which increase transparency by enabling stakeholders to understand and contest automated type of decisions [19]. Table 20.2 summarizes some ethical principles and their operational limitations in AI system design.

Embedding ethics throughout the AI lifecycle—from data collection to deployment—supports responsible innovation and sustainable adoption.

20.3 Legal and Regulatory Frameworks

AI systems working with sensitive fields are highly regulated by legal governance to ensure they do not violate laws and regulations, which is very

TABLE 20.2: Ethical Principles and Design Implications in AI Systems

Principle	Objective	Strategy
Fairness	Equity	Auditing
Transparency	Explainability	Visualization
Accountability	Responsibility	Oversight
Privacy	Protection	Encryption
Beneficence	Welfare	Assessment
Robustness	Stability	Testing
Inclusiveness	Diversity	Representation
Security	Protection	Hardening
Reliability	Consistency	Validation
Sustainability	Efficiency	Optimization
Consent	Autonomy	Disclosure
Traceability	Accountability	Logging
Adaptability	Flexibility	Updating
Proportionality	Balance	Regulation
Verifiability	Trust	Certification
TransparencyPlus	Clarity	Reporting
Explainability	Understanding	XAI
Resilience	Recovery	Redundancy
AccountabilityPlus	Liability	Governance
Inclusivity	Access	Design
Humanity	Dignity	Safeguards
Neutrality	Impartiality	Calibration
Auditability	Verification	Inspection
Portability	Flexibility	Migration
Interoperability	Compatibility	Standardization

significant to AI systems [18]. Regulatory frameworks have been proposed to solve the data protection, algorithmic accountability and human rights [1] by governments and international organizations through the resolution of these issues. The GDPR, the EU AI Act and industry specific standards of compliance are some of the good and good examples thereof [5]. An automated decision has certain legal issues such as defining liability by securing due process, intellectual property protection and transparency considerations/requirements [15]. The PRCs subject to compliance are increasingly demanding algorithmic checks and records. Table 20.3 shows good legal frameworks to the use of AI in high stakes.

The active integration of regulatory standards in regions is one of the biggest challenges to global AI implementations.

TABLE 20.3: Regulatory Frameworks for AI Governance

Framework	Jurisdiction	Focus
GDPR	Europe	Privacy
EUAIA	Europe	Risk
HIPAA	USA	Health
CCPA	USA	Consumer
ISOIEC	Global	Governance
NIST	USA	Security
OECD	Global	Ethics
UNESCO	Global	Rights
PIPL	China	Data
LGPD	Brazil	Privacy
PDPA	Singapore	Protection
DPDP	India	Regulation
FTC	USA	Enforcement
APEC	Global	Cooperation
EDPB	Europe	Compliance
UKGDPR	UK	Privacy
CCPAPlus	USA	Expansion
CNIS	China	Standards
DGA	Europe	Sharing
DMA	Europe	Competition
DSA	Europe	Platforms
AIHLEG	Europe	Guidance
ICCPR	Global	Liberties
ITU	Global	Standards
WTO	Global	Trade

20.4 Societal Impact, Governance, and Future Directions

Outside of ethical and legal aspects, AI systems are affecting labor markets [28], social organization and structures [27], democratic processes [4] and the attitudes of people towards technology. Automation can alter the employment pattern and algorithmic decision-making influences the availability of social services and finances [12]. Automation can change the employment pattern and algorithmic decision-making to social Some good governance models encourage stakeholder involvement, institutional transparency and social responsibility [14]. Participatory design models and human-in-the-loop systems assist in reducing the automation-democratic balance niche [13]. In-the-loop systems and participatory design models aid in striking a balance between

TABLE 20.4: Societal Impacts and Governance Strategies

Impact	Effect	Governance
Employment	Displacement	Reskilling
Trust	Acceptance	Transparency
Divide	Inequality	Inclusion
Autonomy	Dependency	Supervision
Power	Centralization	Oversight
Privacy	Exposure	Regulation
Security	Vulnerability	Hardening
Creativity	Homogenization	Diversity
Democracy	Manipulation	Safeguards
Culture	Standardization	Preservation
Education	Transformation	Modernization
Healthcare	Automation	Governance
Justice	Bias	Auditing
Economy	Polarization	Redistribution
Environment	Consumption	Sustainability
Innovation	Acceleration	Coordination
Labor	Restructuring	Protection
Media	Distortion	Verification
Communication	Fragmentation	Moderation
Mobility	Disruption	Regulation
Identity	Profiling	Safeguards
Accessibility	Expansion	Enablement
Governance	Complexity	Harmonization
Welfare	Dependency	Support
Globalization	Concentration	Balancing

automation and democracy nulls. Table 20.4 outlines important social effects and governance frameworks that relate to the high-stakes AI systems. Further studies ought to concern interdisciplinary cooperation, participatory forms of governance and mechanisms of ongoing ethical assessment.

20.5 Conclusion

This chapter has discussed the ethical, legal, and social issues that this type of decision system has been linked to the use of AI as a deployment system of high stake decisions. Analysis it has carried out through a survey has revealed some of the most vital risks, regulatory frameworks and governance approach to responsible AI adoption. In ensuring trustful AI it is necessary to take ethical reasoning, legal compliance and social accountability into the technical design. The further partnership between certain researchers, policymakers and practitioners will be highly critical in streamlining the AI innovation with human values and societal well-being.

Bibliography

[1] John Babikian. Securing rights: Legal frameworks for privacy and data protection in the digital era. *Law Research Journal*, 1(2):91–101, 2023.

[2] Shreya Chappidi, Jennifer Cobbe, Chris Norval, Anjali Mazumder, and Jatinder Singh. Accountability capture: How record-keeping to support ai transparency and accountability (re)shapes algorithmic oversight. *arXiv preprint arXiv:2510.04609*, 2025.

[3] Ben Chester Cheong. Transparency and accountability in ai systems: safeguarding wellbeing in the age of algorithmic decision-making. *Frontiers in Human Dynamics*, 6:1421273, 2024.

[4] Mark Coeckelbergh. Artificial intelligence, the common good, and the democratic deficit in ai governance. *AI and Ethics*, 5(2):1491–1497, 2025.

[5] Martin Ebers. Truly risk-based regulation of artificial intelligence how to implement the eu's ai act. *European Journal of Risk Regulation*, pages 1–20, 2024.

[6] Lawrence Emma. Explainable ai for high-stakes decision making in healthcare. 2024.

[7] Vijay Govindarajan, Pratik Patel, Sahil Tripathi, Md Azizul Hoque, and Gautam Siddharth Kashyap. Magic-enhanced keyword prompting for zero-shot audio captioning with clip models. *arXiv preprint arXiv:2509.12591*, 2025.

[8] Ansarullah Hasas, Musawer Hakimi, Amir Kror Shahidzay, and Abdul Wajid Fazil. Ai for social good: Leveraging artificial intelligence for community development. *Journal of Community Service and Society Empowerment*, 2(02):196–210, 2024.

[9] L. Judijanto, A. Mudinillah, R. Rahman, and N. Joshi. Ai and social equity: Challenges and opportunities in the age of automation. *Journal of Social Science Utilizing Technology*, 3(1):42–51, 2025.

[10] Ms Zaiba Khan. *Introduction To Artificial Intelligence And Human-centric Computing*. OrangeBooks Publication, 2024.

[11] D. Khazanchi. Navigating digital human rights in the age of ai. *Ethics and Information Technology*, 2025. Published online 2025.

[12] Hariharan Pappil Kothandapani. Social implications of algorithmic decision-making in housing finance: Examining the broader social impacts of deploying machine learning in lending decisions, including potential disparities and community effects. *Journal of Knowledge Learning and Science Technology ISSN: 2959-6386 (online)*, 4(1):78–97, 2025.

[13] Sushant Kumar, Sumit Datta, Vishakha Singh, Deepanwita Datta, Sanjay Kumar Singh, and Ritesh Sharma. Applications, challenges, and future directions of human-in-the-loop learning. *IEEE Access*, 12:75735–75760, 2024.

[14] CHANGKUI LI. Ai-driven governance: Enhancing transparency and accountability in public administration. *Digital Society & Virtual Governance*, 1(1):1–16, 2025.

[15] Brady Lund, Zeynep Orhan, Nishith Reddy Mannuru, Ravi Varma Kumar Bevara, Brett Porter, Meka Kasi Vinaih, and Padmapadanand Bhaskara. Standards, frameworks, and legislation for artificial intelligence (ai) transparency. *AI and Ethics*, pages 1–17, 2025.

[16] David Mark, Tomás McInerney, and John Morison. Regulating automated decision-making in the justice system: what is the problem? *Handbook on Public Policy and Artificial Intelligence*, pages 339–352, 2024.

[17] David K McGraw. Ethical responsibility in the design of artificial intelligence (ai) systems. *International Journal on Responsibility*, 7(1):4, 2024.

[18] Madaoui Nadjia. The impact of artificial intelligence on legal systems: challenges and opportunities. *Проблеми законності*, (164):285–303, 2024.

[19] Evandro S Ortigossa, Thales Gonçalves, and Luis Gustavo Nonato. Explainable artificial intelligence (xai)—from theory to methods and applications. *IEEE Access*, 12:80799–80846, 2024.

[20] Otavio Parraga, Martin D More, Christian M Oliveira, Nathan S Gavenski, Lucas S Kupssinskü, Adilson Medronha, Luis V Moura, Gabriel S Simões, and Rodrigo C Barros. Fairness in deep learning: A survey on vision and language research. *ACM Computing Surveys*, 57(6):1–40, 2025.

[21] Prasanna Kumar Perugu. Ethical data governance: Strategies for reducing bias in enterprise systems. *Available at SSRN 5188642*, 2024.

[22] Katia Schwerzmann and Alexander Campolo. "desired behaviors": alignment and the emergence of a machine learning ethics. *AI & SOCIETY*, pages 1–14, 2025.

[23] A Sharma, R Miller, and L Tan. Explainable ai models for high-stakes decision-making in finance. *Eureka Journal of Artificial Intelligence and Data Innovation*, 1(1):7–16, 2025.

[24] Hussein Kamaldeen Smith. Autonomous ai governance and ethical decision-making frameworks for high-stakes systems in smart cities, fintech, and telecom ecosystems. 2025.

[25] Arpita Soni, Sahil Tripathi, Gautam Siddharth Kashyap, Manaswi Kulahara, Mohammad Anas Azeez, Zohaib Hasan Siddiqui, Nipun Joshi, and Jiechao Gao. Can we predict your next move without breaking your privacy? *arXiv preprint arXiv:2507.08843*, 2025.

[26] D. Vieru and R.-M. Schmitt. Transparency and explainable ai: Bridging privacy, fairness, and accountability. *Review of International Comparative Management*, 26(5):994–1012, 2025.

[27] Qin Weng, Jie Ren, Tom Mattson, Xuefei Deng, and KD Joshi. Rewriting the margins: Can artificial intelligence agents reshape the power dynamics in social media? *ACM SIGMIS Database: the DATABASE for Advances in Information Systems*, 56(3):7–12, 2025.

[28] Zub Mariya Ya. Transformation of labor market infrastructure under the influence of artificial intelligence. *Бизнес Информ*, (8 (511)):146–153, 2020.

[29] Oben Yapar. Explainable ai in national security: Enhancing trust and accountability. *National Security: Enhancing Trust and Accountability (December 13, 2023). International Journal of Emerging Technologies and Innovative Research (www. jetir. org/ UGC and issn Approved), ISSN*, pages 2349–5162, 2023.

Chapter 21

Cross-Sector Lessons and Roadmap: Toward Trustworthy and Domain-Aware AI

Abdullah Mohammad

21.1 Introduction

The rapid growth of Artificial Intelligence (AI) across domains like healthcare [5], finance [10], transportation [6], governance and cybersecurity [14] has showed both its transformative potential and its associated risks. While domain-specific solutions have achieved important success, recurring challenges related to trustworthiness [20], transparency [12], fairness [11] and performance [18] continue to limit any type of large-scale and long-term deployment. Cross-sector analysis provides valuable data into common patterns, best practices, and systemic weaknesses in AI system design and governance. Understanding these shared experiences enables the development of generalized frameworks for responsible and domain-aware intelligence. Table 21.1 summarizes key challenges and lessons observed across major application sectors.

This chapter concludes cross-sector experiences in Section 21.2 presents combined roadmap for trustworthy AI development in Section 21.3, discusses implementation strategies and evaluation mechanisms in Section 21.4 and concludes with future directions in Section 21.5.

21.2 Cross-Sector Lessons for Trustworthy AI

Despite differences in operational environments and stakeholder expectations [8] there are some common lessons grow from cross-domain AI deployments. First, transparency and explainability are quite important for building institutional and public trust [19]. Systems lacking interpretability mainly

DOI: 10.1201/9781003774815-21

TABLE 21.1: Cross-Sector Challenges and Key Lessons in AI Deployment

Sector	Challenge	Lesson
Healthcare	Transparency	Oversight
Finance	Compliance	Explainability
Transportation	Safety	Validation
Cybersecurity	Attacks	Monitoring
Governance	Trust	Participation
Education	Accessibility	Inclusion
Manufacturing	Automation	Optimization
Retail	Personalization	Privacy
Energy	Reliability	Resilience
Agriculture	Forecasting	Adaptation
Telecom	Latency	Efficiency
Defense	Security	Control
Logistics	Coordination	Integration
Media	Misinformation	Verification
Insurance	Risk	Transparency
Banking	Fraud	Detection
Aviation	Precision	Redundancy
Pharmaceuticals	Regulation	Compliance
Construction	Planning	Optimization
Tourism	Volatility	Diversification
RealEstate	Valuation	Analytics
Ecommerce	Scalability	Automation
Mining	Safety	Monitoring
Sports	Analytics	Performance
Entertainment	Piracy	Protection

face resistance from regulators and end-users. Second fairness [15] and bias mitigation [17] must be used throughout the system lifecycle. Many failures in real-world deployments comes from unrepresentative data and inadequate evaluation protocols. Third, continuous monitoring [9] and adaptive learning [7] mechanisms are important to address evolving risks and distributional shifts. Table 21.2 shows recurring design principles and governance practices adopted across sectors.

These lessons focuses on that trustworthy AI is not achieved through isolated technical solutions, but through coordinated technical, organizational and regulatory efforts.

21.3 Integrated Roadmap for Domain-Aware AI Development

Building on cross-sector data it is combined roadmap for domain-aware and trustworthy AI must address the full system lifecycle [4]. This roadmap does have four major phases: design, development, deployment and governance.

TABLE 21.2: Common Design Principles for Trustworthy AI Systems

Principle	Objective	Adoption
Explainability	Transparency	High
Fairness	Equity	High
HumanLoop	Accountability	High
Privacy	Protection	High
Robustness	Reliability	Medium
Security	Defense	High
Scalability	Expansion	Medium
Interpretability	Clarity	Medium
Resilience	Stability	Medium
Governance	Compliance	High
Reproducibility	Consistency	Medium
Auditability	Verification	High
Portability	Flexibility	Medium
Sustainability	Efficiency	Medium
Interoperability	Compatibility	High
Transparency	Trust	High
Adaptability	Evolution	Medium
Reliability	Continuity	High
Traceability	Tracking	Medium
Inclusivity	Accessibility	Medium
Modularity	Maintainability	Medium
Extensibility	Growth	Medium
Observability	Visibility	Medium
Configurability	Customization	Medium
Standardization	Uniformity	High

During the design phase, stakeholder engagement and ethical impact assessments are important. The development phase focuses on data governance, model validation and security testing. Deployment requires scalable monitoring infrastructures and explainability interfaces. Governance mechanisms which secure regulatory compliance, auditing and continuous improvement [13]. Table 21.3 shows a lifecycle-based roadmap for developing domain-aware AI systems.

21.4 Implementation Strategies and Evaluation Frameworks

Translating conceptual roadmaps into operational systems which requires concrete implementation strategies and standardized evaluation frameworks [2]. Successful cross-sector initiatives that focuses on modular architectures [3], interoperability [1] and continuous feedback mechanisms [3]. Hybrid human-AI

TABLE 21.3: Lifecycle Roadmap for Trustworthy and Domain-Aware AI

Phase	**Activities**	**Outcome**
Design	Ethics	Mitigation
Development	Validation	Reliability
Deployment	Monitoring	Stability
Operation	Auditing	Compliance
Governance	Alignment	Accountability
DataCollection	Acquisition	Quality
Preprocessing	Cleaning	Consistency
Training	Optimization	Accuracy
Evaluation	Benchmarking	Verification
Security	Hardening	Resilience
Privacy	Anonymization	Protection
Explainability	Interpretation	Transparency
Maintenance	Updating	Continuity
Scaling	Expansion	Efficiency
Retirement	Decommissioning	Closure
Documentation	Recording	Traceability
Versioning	Control	Reproducibility
Testing	Simulation	Robustness
Calibration	Adjustment	Precision
Integration	Synchronization	Compatibility
Customization	Personalization	Usability
Feedback	Collection	Improvement
Localization	Adaptation	Relevance
Certification	Validation	Legitimacy
Archiving	Preservation	Accessibility

decision systems which are increasingly adopted to balance automation with expert judgment [16]. Evaluation frameworks uses technical metrics, ethical indicators and societal effect evaluations. There are some benchmarking initiatives and regulatory sandboxes that further have responsible experimentation. Table 21.4 summarizes implementation strategies and evaluation approaches for trustworthy AI.

Standardized evaluation practices enhance comparability, transparency, and institutional learning across sectors.

21.5 Conclusion

The cross-sector experience is present in this chapter to pinpoint common challenges, design concepts and governance systems of reliable and domain-aware

TABLE 21.4: Implementation and Evaluation Strategies for Trustworthy AI

Strategy	Purpose	Context
Modularity	Scalability	Enterprise
Collaboration	Oversight	Safety
Auditing	Control	Regulation
Benchmarking	Evaluation	Research
Sandboxes	Testing	Policy
Explainability	Transparency	Healthcare
Federation	Privacy	Distribution
BiasDetection	Fairness	Lending
AdversarialTesting	Security	Infrastructure
Uncertainty	Awareness	Decisions
Governance	Integrity	Public
Ethics	Compliance	Policy
Documentation	Traceability	Industry
Versioning	Reproducibility	Platforms
Feedback	Improvement	Consumers
Monitoring	Surveillance	Operations
Encryption	Protection	Data
Authentication	Verification	Access
Simulation	Validation	Development
Redundancy	Reliability	Systems
Standardization	Compatibility	Integration
Certification	Approval	Regulation
Automation	Efficiency	Production
Logging	Accountability	Governance
Resilience	Stability	Infrastructure

AI. It demonstrated the significance of such integrated lifecycle management, interdisciplinary cooperation and constant assessment as well through comparative analysis. The research ought to be conducted in future to reconcile the global governance structures, enhance explainable and robust learning techniques and reinforce human practices of AI. It takes collective knowledge applied to domains to speed up the construction of responsible, resilient and socially aligned AI systems by stakeholders.

Bibliography

[1] Yik Chan Chin, David A Raho, Hag-Min Kim, Chunli Bi, James Ong, Jingbo Huang, and Serge Stinckwich. Interoperability in ai safety governance: Ethics, regulations, and standards. *arXiv preprint arXiv:2601.06153*, 2026.

[2] Moacir Godinho Filho, Sofia Vieira Queiroz de Almeida, Murís Lage Junior, Lauro Osiro, Bruna Lima, and Mario Henrique Callefi. A path to follow to overcome foundational barriers to the adoption of artificial intelligence within the manufacturing industry: a conceptual framework. *Enterprise Information Systems*, 19(1-2):2458685, 2025.

[3] Vijay Govindarajan, Pratik Patel, Sahil Tripathi, Md Azizul Hoque, and Gautam Siddharth Kashyap. Magic-enhanced keyword prompting for zero-shot audio captioning with clip models. *arXiv preprint arXiv:2509.12591*, 2025.

[4] Zayadul Hasan. Iot-driven implementation of ai predictive models for real-time performance enhancement of perovskite and tandem photovoltaic systems. *ASRC Procedia: Global Perspectives in Science and Scholarship*, 1(01):1031–1065, 2025.

[5] Masooma Hassan, Andre Kushniruk, and Elizabeth Borycki. Barriers to and facilitators of artificial intelligence adoption in health care: scoping review. *JMIR Human Factors*, 11:e48633, 2024.

[6] Gulshan Kumar and Ali Altalbe. Artificial intelligence (ai) advancements for transportation security: in-depth insights into electric and aerial vehicle systems. *Environment, Development & Sustainability*, 27(11), 2025.

[7] Suwen Li and Yunzhe Huang. Dynamic distributionally robust optimization with adaptive ambiguity sets: A machine learning-enhanced framework for market decision-making. *IEEE Access*, 2025.

[8] Michela Magas and Dimitris Kiritsis. Industry commons: an ecosystem approach to horizontal enablers for sustainable cross-domain industrial innovation (a positioning paper). *International Journal of Production Research*, 60(2):479–492, 2022.

[9] Fazliddin Makhmudov, Dusmurod Kilichev, Ulugbek Giyosov, and Farkhod Akhmedov. Online machine learning for intrusion detection in electric vehicle charging systems. *Mathematics*, 13(5):712, 2025.

[10] Prasenjit Roy, Biswajit Ghose, Premendra Kumar Singh, Pankaj Kumar Tyagi, and Asokan Vasudevan. Artificial intelligence and finance: A bibliometric review on the trends, influences, and research directions. *F1000Research*, 14:122, 2025.

[11] Adnan Shahid, Adrian Kliks, Ahmed Al-Tahmeesschi, Ahmed Elbakary, Alexandros Nikou, Ali Maatouk, Ali Mokh, Amirreza Kazemi, Antonio De Domenico, Athanasios Karapantelakis, et al. Large-scale ai in telecom: Charting the roadmap for innovation, scalability, and enhanced digital experiences. *arXiv preprint arXiv:2503.04184*, 2025.

[12] Karthik Shivashankar, Ghadi S Al Hajj, and Antonio Martini. Scalability and maintainability challenges and solutions in machine learning: Systematic literature review. *arXiv preprint arXiv:2504.11079*, 2025.

[13] Gowri Shri. Adaptive data governance models using explainable ai. *International Journal of Emerging Trends in Computer Science and Information Technology*, pages 459–468, 2025.

[14] Arpita Soni, Sahil Tripathi, Gautam Siddharth Kashyap, Manaswi Kulahara, Mohammad Anas Azeez, Zohaib Hasan Siddiqui, Nipun Joshi, and Jiechao Gao. Can we predict your next move without breaking your privacy? *arXiv preprint arXiv:2507.08843*, 2025.

[15] Ezekiel Soremekun, Mike Papadakis, Maxime Cordy, and Yves Le Traon. Software fairness: An analysis and survey. *ACM Computing Surveys*, 58(3):1–38, 2025.

[16] Rajarshi Tarafdar. Human-ai collaboration in workflow optimization: A framework for hybrid decision systems in automation-heavy industries. 2025.

[17] Abhinav Tiwari and Hany EZ Farag. Responsible ai framework for autonomous vehicles: Addressing bias and fairness risks. *IEEE Access*, 2025.

[18] Md Palash Uddin, Yong Xiang, Mahmudul Hasan, Jun Bai, Yao Zhao, and Longxiang Gao. A systematic literature review of robust federated learning: Issues, solutions, and future research directions. *ACM Computing Surveys*, 57(10):1–62, 2025.

[19] Jaideep Visave. Transparency in ai for emergency management: Building trust and accountability. *AI and Ethics*, pages 1–14, 2025.

[20] Wenbo Wu and George Konstantinidis. Trust and reputation in data sharing: a survey. *The VLDB Journal*, 2025.

Index

For Product Safety Concerns and Information please contact our
EU representative GPSR@taylorandfrancis.com Taylor & Francis
Verlag GmbH, Kaufingerstraße 24, 80331 München, Germany